AF389150

Laser Technologies in Metal-Based Materials

Laser Technologies in Metal-Based Materials

Editor

Alina A. Manshina

MDPI • Basel • Beijing • Wuhan • Barcelona • Belgrade • Manchester • Tokyo • Cluj • Tianjin

Editor
Alina A. Manshina
Saint Petersburg State
University
Saint Petersburg, Russia

Editorial Office
MDPI
St. Alban-Anlage 66
4052 Basel, Switzerland

This is a reprint of articles from the Special Issue published online in the open access journal *Materials* (ISSN 1996-1944) (available at: https://www.mdpi.com/journal/materials/special_issues/Laser_Technol_Met_Mater).

For citation purposes, cite each article independently as indicated on the article page online and as indicated below:

LastName, A.A.; LastName, B.B.; LastName, C.C. Article Title. *Journal Name* **Year**, *Volume Number*, Page Range.

ISBN 978-3-0365-8314-3 (Hbk)
ISBN 978-3-0365-8315-0 (PDF)

Contents

About the Editor

Alina A. Manshina

Alina Manshina is a professor at the Institute of chemistry, St. Petersburg State University, Russia. Prof. A. Manshina received her Ph.D. in physics and mathematics (specialty 'laser physics') and her Doctor degree in chemistry (specialty 'solid state chemistry'). That is why her scientific interests are in interdisciplinary fields related to laser-induced processes and phenomena. Her current research activities are focused on the laser synthesis of metal and composite nanomaterials for sensing applications, electrochemistry, and biomedicine; laser-switchable bioactive molecules for photopharmacology; and laser additive technologies for 3D ceramics.

Preface to "Laser Technologies in Metal-Based Materials"

The scope of "Laser Technologies in Metal-Based Materials" is laser-induced physicochemical processes related to the modification or/and synthesis of various metal materials and structures. This is an urgent topic motivated by highly intensive and continuous progress in scientific fields, such as laser physics, laser chemistry, and laser material science. Thanks to such a fundamental background, laser-related technologies are currently dominating many modern industries and are often delivering much more efficient solutions compared to other options.

Thus, this Special Issue aims to bring the fields of laser technologies and metal nanostructures together to explore the benefits of both. It should be noted that the laser effect on a material can initiate physical phenomena (heating, phase transitions, etc.) and/or chemical phenomena (oxidation, reduction, chemical transformations). The articles in the current reprint appropriately combine physical and chemical phenomena and offer new advanced laser technologies to modern society. Here, we consider the different aspects of laser technologies for the fabrication of metal-based functional materials, as numerous modern instruments and devices are based on processes related to metal nanostructures.

"Laser Technologies in Metal-Based Materials" is addressed to wide audience, from the community of scientists working in various laser-related research fields to engineers and technology developers. The authors are grateful to MDPI Section Managing Editor Ms. Freya Dong, as without her support, this Special Issue would not have been possible.

Alina A. Manshina
Editor

Editorial

Special Issue "Laser Technologies in Metal-Based Materials"

Alina Manshina

Institute of Chemistry, Saint-Petersburg State University, 26 Universitetskii Prospect, Saint-Petersburg 198504, Russia; a.manshina@spbu.ru

The first publication, analyzing the prospects for the use of laser radiation, was published under the authorship of the American physicist Arthur Shawlow in November 1960 (Schawlow, A. L. Bell Lab. Rec., November, 403 (1960)) immediately after the creation of the first laser by Theodor Meiman on 16 May 1960. Later, Arthur Shawlow received the Nobel Prize. Subsequently, many brilliant scientists (A. Zeveil, V.S. Letokhov, N.V. Karlov, and many others) joined the topic of laser-induced processes, which ensured rapid progress in this area [1–7]. As a result, new directions in chemistry and physics have been formed—laser chemistry and laser physics, which continue to be a dynamically developing science. These laser-related directions consider the fundamental issues of the synthesis/transformation of substances and the problems of high precision and highly controlled laser technologies. Insightful publications of the late 20th century reporting on original ideas of laser irradiation use for various processes of materials transformation and fabrication [8–10] turned into extensive areas related to laser technologies since the beginning of the 21st century.

This Special Issue aims to bring the fields of laser technologies and metal nanostructures together for both benefits. We consider different aspects of laser technologies for fabrication of metal-based functional nanomaterials here, as numerous modern instruments and devices are based on processes related to metal nanostructures. It should be noted that the laser effect on a material can initiate physical phenomena (heating, phase transitions, etc.) and/or chemical phenomena (oxidation, reduction, chemical transformations). Thus, the articles of the current Special issue harmoniously combine physical and chemical phenomena and offer advanced laser technologies to modern society.

Regarding publications in laser-induced physical processes, one can find the article by A. V. Agapovichev et al. on selective laser melting to produce Ni-Cr-Al-Ti-Based Superalloy [11]. The authors of the article present sintering processes by pulsed nanosecond laser for obtaining aerosol agglomerates of Pt, Au, and Ag NPs [12]. The interesting combination of processes of laser-induced surface texturing simultaneously with laser-induced anchoring of silver NPs from colloidal solution is discussed by Jakub Siegel et al. in [13]. Such textured polymer surfaces decorated with Ag NPs can be prospective antimicrobial coatings. Another example of laser-induced physical phenomena is laser shock peening, demonstrating significantly improving the fretting fatigue life of TC11 titanium alloy [14]. In the article by Piotr Kupracz et al. [15] laser re-solidification was demonstrated as an approach for the modulation of morphology and structure of metal-decorated TiO_2 nanotubes to obtain visible light harvesting.

Interesting advanced approaches for creating nanostructured metal materials with various functionality were presented in laser-induced chemical processes. Thus, laser ablation of monocrystalline silicon in isopropanol containing $AgNO_3$ allowed the single-step formation of Ag-decorated Si microspheres with SERS performance [16]. Here, the physical process of laser ablation is accompanied by the chemical process of Ag NPs formation onto ablated Si species. Femtosecond laser reductive sintering allowed for obtaining high-purity Cu patterns from CuO NPs inks [17]. At the same time, a variant of selective laser reductive sintering created copper and nickel microsensors for non-enzymatic glucose detection [18]. Highly controllable decoration of substrates by plasmonic

Citation: Manshina, A. Special Issue "Laser Technologies in Metal-Based Materials". *Materials* **2023**, *16*, 4511. https://doi.org/10.3390/ma16134511

Received: 17 April 2023
Accepted: 16 June 2023
Published: 21 June 2023

Ag, Pt NPs with uniform or periodic NPs distribution was demonstrated due to laser-induced deposition [19]. This laser-induced process is based on the photodecomposition of metal-containing precursors and following redox processes onto the substrate surface. Interestingly, a similar process can be realized as a laser-induced thermal process resulting in composite materials based on iridium, gold, and platinum [20].

Conflicts of Interest: The authors declare no conflict of interest.

References

1. Zewail, A.H. Laser selective chemistry—Is it possible? *Phys. Today* **1980**, *33*, 27. [CrossRef]
2. Letokhov, V.S. Laser-induced chemistry. *Nature* **1983**, *305*, 103. [CrossRef]
3. Karlov, N.V. Laser-induced chemical reactions. *Appl. Opt.* **1974**, *13*, 301. [CrossRef] [PubMed]
4. Geohegan, D.B.; Puretzky, A.A.; Duscher, G.; Pennycook, S.J. Time-resolved imaging of gas phase nanoparticle synthesis by laser ablation. *Appl. Phys. Lett.* **1998**, *72*, 2987. [CrossRef]
5. Semaltianos, N.G. Nanoparticles by Laser Ablation. *Crit. Rev. Solid State Mater. Sci.* **2010**, *35*, 105. [CrossRef]
6. Kim, D.; Jang, D. Synthesis of nanoparticles and suspensions by pulsed laser ablation of microparticles in liquid. *Appl. Surf. Sci.* **2007**, *253*, 8045. [CrossRef]
7. Zeng, H.; Du, X.W.; Singh, S.C.; Kulinich, S.A.; Yang, S.; He, J.; Cai, W. Nanomaterials via laser ablation/irradiation in liquid: A review. *Adv. Funct. Mater.* **2012**, *22*, 1333. [CrossRef]
8. Roberts, M.A.; Rossier, J.S.; Bercier, P.; Girault, H. UV Laser Machined Polymer Substrates for the Development of Microdiagnostic Systems. *Anal. Chem.* **1997**, *69*, 2035. [CrossRef] [PubMed]
9. Nakano, S.; Matsuoka, T.; Kiyama, S.; Kawata, H.; Nakamura, N.; Nakashima, Y.; Tsuda, S.; Nishiwaki, H.; Ohnishi, M.; Nagaoka, I. Laser Patterning Method for Integrated Type a-Si Solar Cell Submodules. *Jpn. J. Appl. Phys.* **1986**, *25*, 1936. [CrossRef]
10. Keicher, D.M.; Smugeresky, J.E. The laser forming of metallic components using particulate materials. *JOM* **1997**, *49*, 51. [CrossRef]
11. Agapovichev, A.V.; Khaimovich, A.I.; Smelov, V.G.; Kokareva, V.V.; Zemlyakov, E.V.; Babkin, K.D.; Kovchik, A.Y. Multiresponse Optimization of Selective Laser Melting Parameters for the Ni-Cr-Al-Ti-Based Superalloy Using Gray Relational Analysis. *Materials* **2023**, *16*, 2088. [CrossRef] [PubMed]
12. Khabarov, K.; Nouraldeen, M.; Tikhonov, S.; Lizunova, A.; Seraya, O.; Filalova, E.; Ivanov, V. Comparison of Aerosol Pt, Au and Ag Nanoparticles Agglomerates Laser Sintering. *Materials* **2022**, *15*, 227. [CrossRef] [PubMed]
13. Siegel, J.; Savenkova, T.; Pryjmaková, J.; Slepička, P.; Šlouf, M.; Švorčík, V. Surface Texturing of Polyethylene Terephthalate Induced by Excimer Laser in Silver Nanoparticle Colloids. *Materials* **2021**, *14*, 3263. [CrossRef] [PubMed]
14. Yang, X.; Zhang, H.; Cui, H.; Wen, C. Effect of Laser Shock Peening on Fretting Fatigue Life of TC11 Titanium Alloy. *Materials* **2020**, *13*, 4711. [CrossRef] [PubMed]
15. Kupracz, P.; Grochowska, K.; Karczewski, J.; Wawrzyniak, J.; Siuzdak, K. The Effect of Laser Re-Solidification on Microstructure and Photo-Electrochemical Properties of Fe-Decorated TiO_2 Nanotubes. *Materials* **2020**, *13*, 4019. [CrossRef] [PubMed]
16. Gurbatov, S.; Puzikov, V.; Modin, E.; Shevlyagin, A.; Gerasimenko, A.; Mitsai, E.; Kulinich, S.A.; Kuchmizhak, A. Ag-Decorated Si Microspheres Produced by Laser Ablation in Liquid: All-in-One Temperature-Feedback SERS-Based Platform for Nanosensing. *Materials* **2022**, *15*, 8091. [CrossRef] [PubMed]
17. Mizoshiri, M.; Yoshidomi, K. Cu Patterning Using Femtosecond Laser Reductive Sintering of CuO Nanoparticles under Inert Gas Injection. *Materials* **2021**, *14*, 3285. [CrossRef] [PubMed]
18. Tumkin, I.I.; Khairullina, E.M.; Panov, M.S.; Yoshidomi, K.; Mizoshiri, M. Copper and Nickel Microsensors Produced by Selective Laser Reductive Sintering for Non-Enzymatic Glucose Detection. *Materials* **2021**, *14*, 2493. [CrossRef] [PubMed]
19. Mamonova, D.V.; Vasileva, A.A.; Petrov, Y.V.; Danilov, D.V.; Kolesnikov, I.E.; Kalinichev, A.A.; Bachmann, J.; Manshina, A.A. Laser-Induced Deposition of Plasmonic Ag and Pt Nanoparticles, and Periodic Arrays. *Materials* **2021**, *14*, 10. [CrossRef] [PubMed]
20. Panov, S.M.; Khairullina, M.E.; Vshivtcev, S.F.; Ryazantsev, N.M.; Tumkin, I.I. Laser-Induced Synthesis of Composite Materials Based on Iridium, Gold and Platinum for Non-Enzymatic Glucose Sensing. *Materials* **2020**, *13*, 3359. [CrossRef] [PubMed]

 *materials*

Article

Multiresponse Optimization of Selective Laser Melting Parameters for the Ni-Cr-Al-Ti-Based Superalloy Using Gray Relational Analysis

Anton V. Agapovichev [1,2,*], Alexander I. Khaimovich [1], Vitaliy G. Smelov [1], Viktoriya V. Kokareva [1], Evgeny V. Zemlyakov [3], Konstantin D. Babkin [3] and Anton Y. Kovchik [3]

[1] Engine Production Technology Department, Samara National Research University, 34 Moskovskoye Shosse, 443086 Samara, Russia
[2] Turbomachinery and Heat Transfer Laboratory, Aerospace Engineering Department, Technion-Israel Institute of Technology, Haifa 3200003, Israel
[3] World-Class Research Center "Advanced Digital Technologies", State Marine Technical University, 190121 Saint Petersburg, Russia
* Correspondence: agapovichev5@mail.ru

Abstract: The selective laser melting technology is of great interest in the aerospace industry since it allows the implementation of more complex part geometries compared to the traditional technologies. This paper presents the results of studies to determine the optimal technological parameters for scanning a Ni-Cr-Al-Ti-based superalloy. However, due to a large number of factors affecting the quality of the parts obtained by selective laser melting technology, the optimization of the technological parameters of the scanning is a difficult task. In this work, the authors made an attempt to optimize the technological scanning parameters which will simultaneously correspond to the maximum values of the mechanical properties ("More is better") and the minimum values of the dimensions of the microstructure defect ("Less is better"). Gray relational analysis was used to find the optimal technological parameters for scanning. Then, the resulting solutions were compared. As a result of the optimization of the technological parameters of the scanning by the gray relational analysis method, it was found that the maximum values of the mechanical properties were achieved simultaneously with the minimum values of the dimensions of a microstructure defect, at a laser power of 250 W and a scanning speed of 1200 mm/s. The authors present the results of the short-term mechanical tests for the uniaxial tension of the cylindrical samples at room temperature.

Keywords: additive manufacturing; selective laser melting; Ni-Cr-Al-Ti; metal powder; mechanical properties; microstructure; gray relational analysis

Citation: Agapovichev, A.V.; Khaimovich, A.I.; Smelov, V.G.; Kokareva, V.V.; Zemlyakov, E.V.; Babkin, K.D.; Kovchik, A.Y. Multiresponse Optimization of Selective Laser Melting Parameters for the Ni-Cr-Al-Ti-Based Superalloy Using Gray Relational Analysis. *Materials* **2023**, *16*, 2088. https://doi.org/10.3390/ma16052088

Academic Editors: Aleksander Muc and Massimiliano Avalle

Received: 30 September 2022
Revised: 7 February 2023
Accepted: 26 February 2023
Published: 3 March 2023

1. Introduction

The current pace of the aviation and rocket and space industry development requires the use of production technologies that make it possible to obtain high-quality products in the shortest possible time at the lowest cost and with minimal post-processing [1,2]. The use of additive manufacturing (AM) can significantly reduce the duration of the technological preparation for the production of new products with a rather complex shape, use fundamentally new design and technological solutions, and ultimately reduce the labor intensity and the cost of manufacturing the products [3]. AM is developing at a faster pace and, above all, processes involving 3D printing of metals and alloys, because they allow the manufacture of geometrically complex components that cannot be easily formed by other traditional processes, and, in addition, minimize the loss of raw materials [4,5].

Ni-based superalloys are the most extensively investigated superalloys in the AM field and have become the materials of choice for fabricating the components destined for high-temperature industrial applications, such as turbine discs and blades [6]. The most commonly processed AM Ni-based superalloys are Inconel 718 and Inconel 625, fabricated with a relative density close to 100% due to their good weldability [7].

Nevertheless, Inconel 718 and Inconel 625 can be used up to around 650 °C for applications under a load. In fact, it is well known that the operative temperature of these alloys is limited under a high load due to the coarsening and the transformation of the metastable γ'' phase into the δ phase, which can significantly reduce the mechanical characteristics [8]. In this regard, the need for alloys operating at temperatures above 650 °C causes increased interest in the development of nickel-based alloys by the AM. Hastelloy X, Inconel 738LC and CM247LC are among these superalloys [7]. As recently reported, a high concentration of Si and C may play a critical role in increasing the cracking susceptibility of Hastel-loy X [9]. The 'Weldability' diagram for a range of nickel-based superalloys (Figure 1) shows that the CM247LC and Inconel 738LC alloys have been considered poor [10].

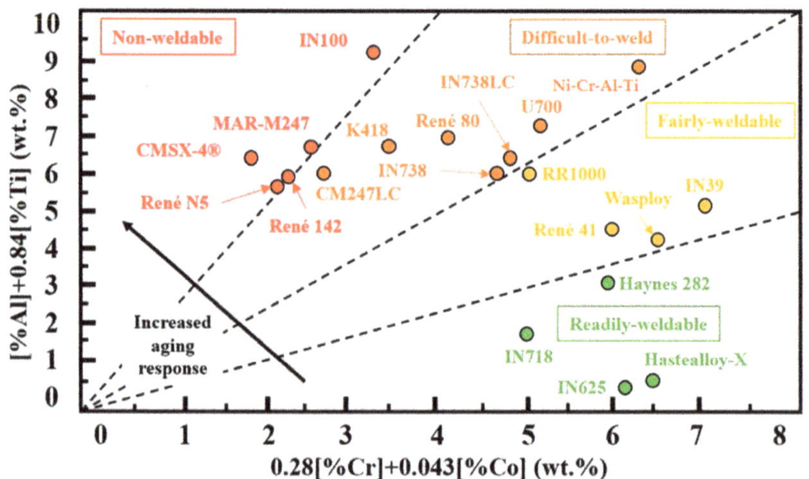

Figure 1. Weldability assessment for nickel-based superalloys [10].

Here, the AM of a Ni-20.5Cr-6Al-3.5Ti (wt%) model alloy is studied. This provides a model Ni-base superalloy with a high Al + Ti content and a high γ' fraction and solvus temperature. High creep resistance, corrosion, oxidation resistance, as well as microstructure stability up to around 850 °C are presented by Ni-Cr-Al-Ti superalloys [11].

Selective laser melting (SLM) is one of the emerging AM technologies that use high-power lasers to create three-dimensional physical objects by fusing metal powders [12,13]. The quality of parts manufactured by SLM technology is affected by a large number of technological parameters [14,15]. By properly understanding and managing these parameters, it is possible to obtain parts of a quality that is not inferior to that of the parts obtained by the traditional production methods. The complexity of the SLM process lies in many thermal, physical, and mechanical interactions and the influence of a large number of technological parameters on them. For example, the SLM Solutions build processor used on the SLM 280HL contains over 100 configurable parameters. [15,16]. There are four groups of technological parameters: laser parameters, scan options, material parameters, and atmospheric parameters [17].

From the literature sources [18,19], it is known that the main technological parameters of scanning are laser power P, (W); scanning speed V, (mm/s); hatch distance h, (μm); layer thickness t, (μm) and scanning strategy type [15]. The technological scanning parameters are often combined into one parameter, which is called volume energy density (VED, J/mm^3) and is determined by Equation (1). This parameter has no physical meaning, but it is used to compare the physical and mechanical properties of a material synthesized at different technological parameters [20].

$$VED = \frac{P}{V \cdot h \cdot t},\tag{1}$$

where P is the laser power, W; V is the scanning speed, mm/s; h is the hatch distance, mm; and t is the layer thickness.

In some papers, the technological parameters of scanning, as a result of which a material with a density close to 100% is synthesized, are called optimal technological parameters [21,22]. To achieve a material density close to 100%, it is necessary to ensure that sufficient heat is supplied to a certain volume of material in order to ensure its complete penetration. As can be seen in Equation (1), this condition can be met by applying various combinations of technological scanning parameters. The numerous combinations of technological scanning parameters, as a result of which a material with a density close to 100% is synthesized, are called the range of optimal technological parameters of scanning (process window). For example, Gong et al. determined the process window for the Ti-6Al-4V titanium alloy [23]. The process window is divided into four areas, and a material with a density close to 100% can be synthesized using one of the combinations of the laser power and the scanning speed. In the work of Vrancken B. [24] on the process window for the SLM of AlSi10Mg and in the work of Montgomery C. et al. [25] on the process window for the SLM of Ti6Al4V studies of the data were carried out on the shape of the melt pools formed as a result of exposure to technological scanning parameters.

However, the quality of the material synthesized by SLM technology is characterized by more than one experiment response characteristic, which can weakly correlate with each other [1]. Therefore, when we faced the problem of simultaneously optimizing several quality characteristics (yield strength, tensile strength, relative elongation, area of non-melting points, pore diameter, etc.), it became clear that it was necessary to find a compromise between their optimal states, depending on the importance of one quality parameter compared to the others.

Gray relational analysis (GRA) is a popular type of optimization method which is used for solving of multi-criteria problems [26,27]. In gray relational analysis, the dimensions of the factors considered are usually different, and their magnitude difference is large. Therefore, the original data are normalized to make the magnitude of the original data of the order of one and to make them dimensionless [28]. Therefore, in order to search for the optimal technological parameters of scanning, the initial experimental data are first normalized in the range from 0 to 1 in accordance with the principles "More is better" for the values of the mechanical properties (target value 1) and "Less is better" for the sizes of the microstructure defects (target value 0). The overall evaluation of the multiple quality characteristics is based on the grey relational grade [29]. Firstly, the gray relational coefficient is determined to quantify the relationship between ideal and actual experimental results. Then, to summarize all the quality characteristics, the grey relational grade is calculated by averaging the gray relational coefficients. The optimal level of the process parameters is the level with the highest grey relational grade [26,28].

2. Materials and Methods

In this paper, a study to determine the optimal technological scanning parameters for the Ni-Cr-Al-Ti-based superalloy to be processed by SLM was conducted based on the gray relational analysis. In the gray relational analysis, the raw experimental data are normalized at first for the reasons discussed earlier. In this study, a linear normalization of the experimental results is performed in the range of 0 to 1, which is called the gray relational generating.

The normalized results for the "more is better" quality characteristic can be expressed as:

$$x_{ij} = \frac{y_{ij} - \min_j y_{ij}}{\max_j y_{ij} - \min_j y_{ij}},$$
(2)

For the "less is better" quality characteristic, the normalized results can be expressed as:

$$x_{ij} = \frac{\max_j y_{ij} - y_{ij}}{\max_j y_{ij} - \min_j y_{ij}},$$
(3)

where y_{ij} is the value of the quality parameter j for the i-th combination of the scanning parameters, and $y_{ij} = \max_n y_{ij}^n$ is the most negative quality characteristic among the $n = 1, \ldots, 9$ combinations of scanning parameters.

The gray relational coefficients ξ_{ij}, which are calculated to determine the relationship between the ideal and the actual results of an experiment, can be expressed as [28–30]:

$$\xi_{ij} = \frac{\min_i \min_j \left| x_j^0 - x_{ij} \right| + \zeta \max_i \max_j \left| x_j^0 - x_{ij} \right|}{\left| x_j^0 - x_{ij} \right| + \zeta \max_i \max_j \left| x_j^0 - x_{ij} \right|},$$
(4)

where x_i^0 is the ideal normalized result (i.e., best normalized result) for the j-th quality characteristics, and $\zeta = [0,1]$ is a distinguishing coefficient, the purpose of which is to weaken the effect of $\max_i \max_j \left| x_j^0 - x_{ij} \right|$ when it gets too big and thus enlarges the difference significance of the relational coefficient. In general, it is set to 0.5 if all the process parameters have equal weighting [29].

Based on the experience of using gray relational analysis to search for rational SLM process parameters, the recommended values of the coefficient ζ are given in Table 1. The same table shows the normalized desired quality characteristics (x_i^0).

Table 1. Constants for conducting gray relational analysis.

Yield Stress, $\sigma_{0.2}$, MPa	Tensile Strength, σ_B, MPa	Relative Elongation, δ, %	Microstructure Desirability Rank	Non-Melting Area, mm^2	Pore Diameter, mm
$\zeta = 0.5$	$\zeta = 0.7$	$\zeta = 0.5$	$\zeta = 0.5$	$\zeta = 0.5$	$\zeta = 0.5$
$x_i^0 = 1.2$	$x_i^0 = 1.2$	$x_i^0 = 1.2$	$x_i^0 = 1$	$x_i^0 = 1$	$x_i^0 = 1$

After obtaining the gray relational coefficients, a weighting method is used to integrate the gray relational coefficients of each experiment into the gray relational grade. The quality characteristics are based on the gray relational grade and can be expressed as [29,30]:

$$\gamma_i = \frac{1}{m} \sum_{j=1}^{m} \xi_{ij},$$
(5)

where m is the number of quality characteristics.

In calculating the gray relational grades, the weighting ratio for the quality characteristics are set as 1:1, i.e., each characteristic has equal importance or relative weighting.

The Ni-Cr-Al-Ti-based superalloy metal powder was used as an initial material. Due to the scanning with the electron microscope Tescan Vega and Energy Dispersive X-rays Spectrometer INCAx-act it was possible to investigate the morphology of surface powder and chemical composition.

The following factors were chosen as variable factors: laser power (P), scanning speed (V), and hatch distance (h). The research by Anthony De Luca et al. shows that the highest density of the Ni-Cr-Al-Ti superalloy achieved is achieved at a VED of 66.6 J/mm^3, and the lowest is at a VED of 44 J/mm^3 [31]. However, a publication by Gokhan Dursun et al. states that there is no relationship between the material density and VED [32]. Therefore, the factors varied in the following ranges (process window from min. to max.): laser power 250, ... , 350 W, scanning speed 600, ... , 1200 mm/s, and hatch distance 0.09, ... , 0.15 mm, which correspond to a VED of 34.7, ... , 97.2 J/mm^3. The samples were prepared at a layer thickness of 0.05 mm; the scanning strategy was zigzag.

Samples were fabricated from Ni-Cr-Al-Ti based superalloy powder on a SLM 280HL 3D printer (SLM Solutions Group AG, Lubeck, Germany). This 3D printer operates using SLM technology, the essence of which is laying the powder manufacturing a part by fusing layers together [6].

The optimal scanning parameters determination was implemented on proportional flat samples with dimensions (L × W × H) 70 × 2 × 15 mm. The study was accomplished according to the fractional factorial design 3 (3-1) with 3 times repetition for each experiment. The experimental design is demonstrated in Table 2.

Table 2. The experimental plan.

Experiment Number	VED, J/mm^3	P, W	V, mm/c	h, mm	t, mm
1	66.7	300	600	0.15	0.05
2	37.0	250	900	0.15	0.05
3	97.2	350	600	0.12	0.05
4	34.7	250	1200	0.12	0.05
5	92.6	250	600	0.09	0.05
6	86.4	350	900	0.09	0.05
7	55.6	300	1200	0.09	0.05
8	55.6	300	900	0.12	0.05
9	38.9	350	1200	0.15	0.05

The mechanical test of the sintered specimens was realized on Instron 8802.

To appoint the non-melting area and the diameter of the pores, thin sections of the cross section of the samples were arranged. Etching of the samples was achieved by electrolytic method at room temperature for 5...10 s in an electrolyte of the following composition: 10 g of citric acid + 10 g of ammonium chloride + 1 L of water.

Microanalysis was implemented on an optical microscope METAM LV-41 in a bright field with ×50...200 magnification. The processing of the acquired images of the microstructure was accomplished in a specialized software product SIAMS.

3. Results and Discussion

3.1. Powder Distribution

The general view of the powder particles is shown in Figure 2. The electron microscope analysis showed that the powder particles mainly have a spherical shape (92%), which is typical for the method of obtaining powders by melt dispersion [33]. The particle size varies in the range of 15–45 microns. The plus and minus fractions are 0.5% and 1.5%, respectively.

The microspectral analysis of the powder particles showed that the chemical composition of the Ni-Cr-Al-Ti-based superalloy powder complies with the manufacturer's quality certificate (TU 1809104096), except for the Al content. The chemical composition of the powder is presented in Table 3.

Table 3. Chemical composition of Ni-Cr-Al-Ti superalloy powder.

Spectrum	Al	Ti	Cr	Co	Ni	Mo	W	Nb	Total
Measured val.	6.08 ± 0.61	3.47 ± 0.12	20.56 ± 0.13	10.43 ± 0.14	54.79 ± 0.33	0.56 ± 0.34	4.11 ± 0.29	-	100.00
TU 1809104096	2.1 – 2.9	3.1 – 3.9	20.0 – 21.8	10.3 – 12.0	base	0.3 – 0.9	3 – 4	0.15 – 0.35	100.00

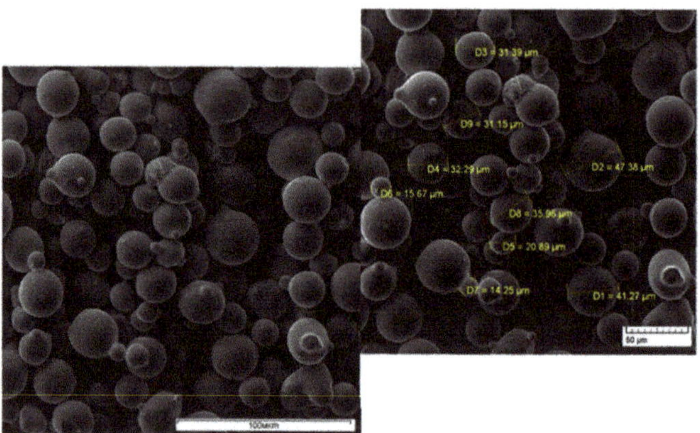

Figure 2. Particles of Ni-Cr-Al-Ti-based superalloy powder.

3.2. Microstructural Characterization and Mechanical Property Tests

During macroanalysis, it was found that on the surface of all the samples there were pores less than 0.1 mm in size. Large defects (non-melting area) of up to 0.6 mm in size were found on sample Nos. 7 and 9 (Figure 3).

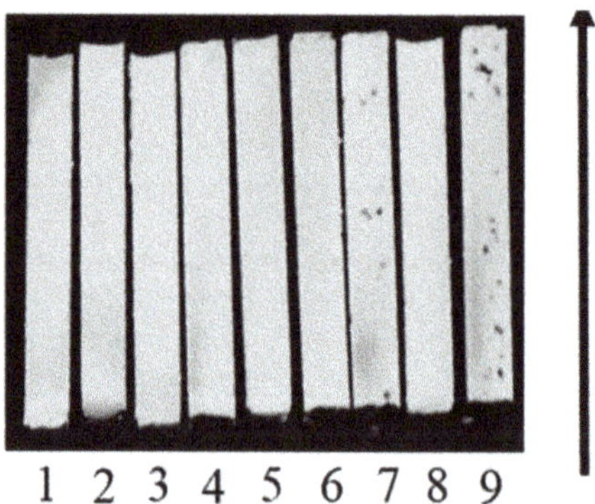

Figure 3. Macrostructure of samples.

In the central part of all the samples, except for sample Nos. 7 and 9, there were single pores (up to 0.03 mm) and non-melting area with a maximum size of 0.1 × 0.25 mm (Figure 4). In the structure of sample Nos. 7 and 9, multiple non-melting spots of up to 0.3 × 0.6 mm in size were observed.

According to the literature, the metal melting is classified into two modes: the conduction mode and the mode with the formation of a penetration channel (melting with deep penetration) [34]. According to the results of the studies by the authors W.E. King and others [35] and E.W. Eagle and others [36], the cross-section of the melt pool formed in the conduction mode is approximately semicircular. Figure 5 shows that the use of the modes with a VED from 34.7 to 66.7 J/mm^3 leads to the formation of a melt pool which is close to semicircular. An increase in VED leads to the transition of the melting of the material to the formation of a penetration channel. An increase in VED to 97.2 J/mm^3 leads to the melting of the material in the deep penetration mode and the formation of keyhole defects. For further analysis and the establishment of a correlation between the technological parameters and the characteristics of the structure, it was necessary to convert the descriptive characteristics of the microstructure into quantitative ones.

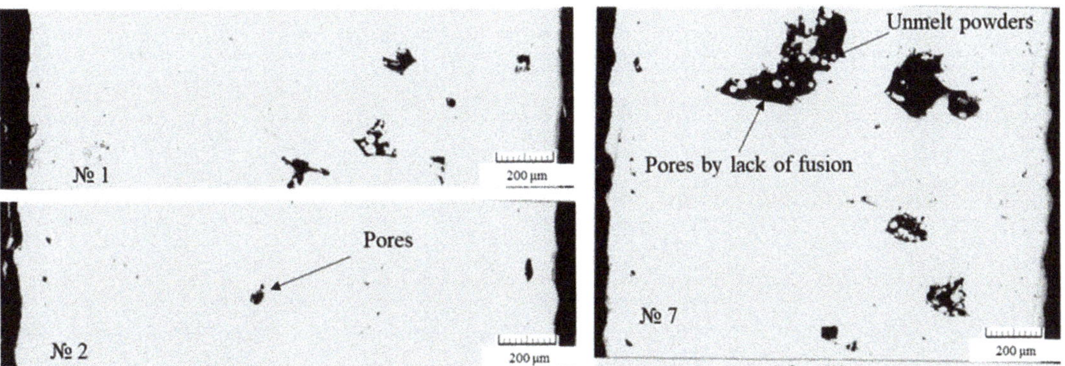

Figure 4. Defects in samples.

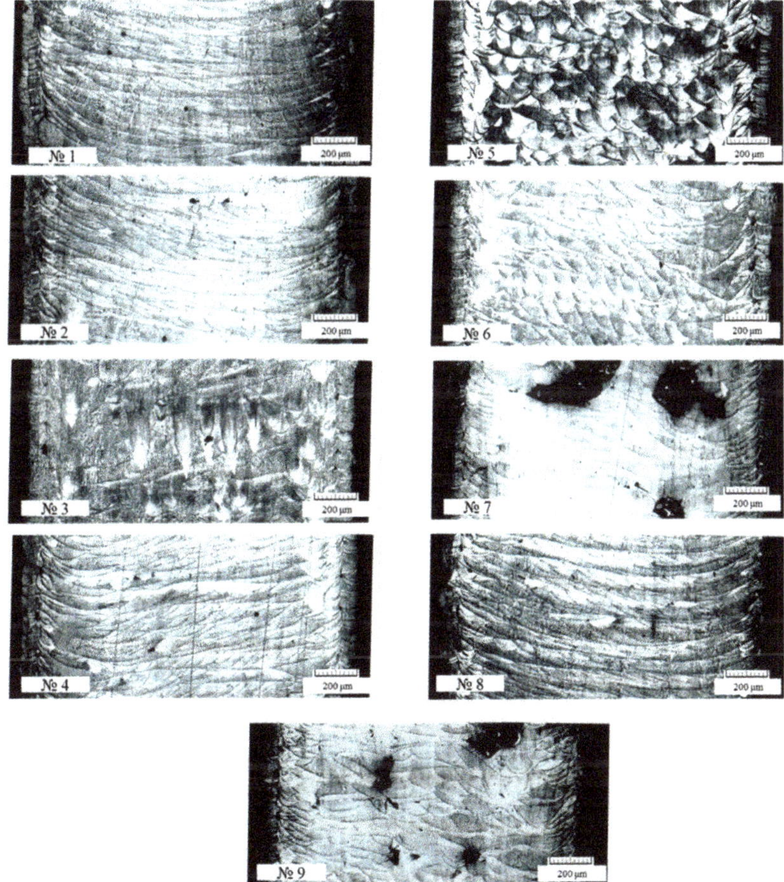

Figure 5. Cross-section comparison of 9 different types of melt pool geometries observed.

For this purpose, the rank of the desirability of the characteristics of the structure of the resulting material was assessed based on the data in Table 4.

Table 4. Rank assessment of a qualitative characteristic.

Structure Characteristic	Quantitative/Qualitative Characteristic		Desirability Rank "Less Is Better"
No penetration between layers	Significant areas of non-penetration	None observed	1–2 9–10
Pores	Single large pores more than 20 microns	Less than 3 in 500 mm^2 More than 3 inclusive in 500 mm^2	6–8 8–10
Structure type	Formation of a molten pool of the "ripple" type—the ratio length/height is not more than 1.8 Elongated grains—length/height ratio more than 1.8		1–4 5–8

In comparison to the as-casted samples, Table 5 reveals that the SLM-manufactured Ni-Cr-Al-Ti superalloy had high tensile strength, much higher than that of the as-casted samples (768 MPa [37]). The highest tensile strength achieved was 949 MPa (300 W and 600 mm/s, equivalent to 66.7 J/mm^3), and the lowest was 774.7 MPa (250 W and 900 mm/s, equivalent to 37 J/mm^3). The SLM-manufactured Ni-Cr-Al-Ti superalloy had high relative elongation, much higher than that of the as-casted samples (3.3 % [37]). The highest relative elongation achieved was 10.2% (250 W and 1200 mm/s, equivalent to 34.7 J/mm^3), and the lowest was 5.2% (250 W and 900 mm/s, equivalent to 37 J/mm^3)

Table 5. Experimental results.

Experiment Number	Yield Stress $\sigma_{0.2}$, MPa	Tensile Strength σ_B, MPa	Relative Elongation δ, %	Microstructure Desirability Rank	Non-Melting Area, mm^2		Pore Diameter, mm
					In the Central Part	Along the Edges	In the Central Part
1	797.6	949.0	8.4	6	0.025	0.008	0.025
2	699.5	774.7	5.2	6	0.002	0.00125	0.025
3	680.8	835.5	7.2	8	0.0014	0.0027	0.03
4	709.9	908.4	10.2	6	0.0008	0.0032	0.023
5	704.0	843.0	7.5	7	0.0006	0.00405	0.03
6	699.1	874.1	8.4	7	0.008	0.01008	0.024
7	680.8	808.3	6.7	10	0.18	0.0021	0.022
8	715.5	880.3	9.1	6	0.0006	0.005	0.02
9	668.4	780.5	5.8	10	0.132	0.0035	0.015

The SLM-manufactured Ni-Cr-Al-Ti superalloy had yield stress elongation, slightly less than that of the as-casted samples (703 MPa % [37]), on samples 2, 3, 6, 7, and 9. In general, the data presented in Table 5 indicate a large variability of the process, which requires the use of additional analysis to select the optimal fusion mode.

To determine the method for finding the optimal scanning parameters, a correlation analysis of the experimental results was carried out; it determined the statistical significance of the relationship between the technological parameters (factors) and each response of the experiment (quality characteristic). If the relationship is significant then Spearman's correlation coefficient will be greater than 0.7, otherwise the relationship is considered insignificant. In the case of a significant relationship between the factors and the experiment responses, it is possible to establish a regression dependence, which will have sufficient reliability in terms of the totality of the characteristics, including the coefficient of determination and Fisher's F-test.

If the correlation coefficients do not confirm the statistical significance of the relationships between the factors and the responses (either all or some of them), then the gray relational analysis method can be used to select the optimal scanning parameters among the nine experiments performed.

The correlation matrix of the factors and characteristics of the experiment response, as presented in Table 6, shows that the range of changes in the scanning parameters in the coefficients of the correlation matrix is not statistically significant, with the exception of the relationship between the pore diameters, the energy density, and the scanning speed.

In addition, the mechanical properties have a weak correlation with the rank assessment of the microstructure characteristics. Thus, in agreement with the statement that in the conditions of the significant variability of the process responses, the analysis methods with fuzzy boundaries of target indicators, such as that of gray relational analysis, should be used. So, according to the combination of features, it was advisable to apply the method of gray relational analysis to determine the optimal scanning parameters for the Ni-Cr-Al-Ti-based superalloy powder. An analysis of the influence of VED and scanning speed V on the pore diameter (Table 6) based on the values of the correlation coefficients equal to 0.71002 and -0.77029, respectively, shows that excess fusion energy increases the tendency to form large pores, and a higher speed of movement of the melt pool promotes their reduction. The influence of these two factors has an almost equal effect. Thus, according to the correlation analysis data, it is clear that the optimal technological parameters for the series of experiments performed lie in the region of lower VED values and higher scanning speed values. This result is justified by the fact that the increasing of the fusion energy leads to the formation of solidification cracks, which is related to the keyhole effect (Figure 5—№3) as well as the pores; the latter is due to the evaporation of small powder particles. However, on the other hand, a lack of energy leads to the appearance of non-melting areas (Figure 4—image on the right; Figure 5—№7). For a more detailed analysis and selection of the optimal mode for growing samples, a gray relational analysis was carried out.

Table 6. Parameter correlation matrix.

Factors/ Experiment Response Characteristics	Yield Stress, $\sigma_{0.2}$, MPa	Tensile Strength, σ_B, MPa	Relative Elongation, δ, %	Microstructure Desirability Rank	Non-Melting Area, mm^2		Pore Diameter, mm
					In the Central Part	Along the Edges	In the Central Part
VED, J/mm^3	0.03822	0.19419	0.10204	0.46347	-0.31511	0.40573	**0.71002**
P, W	-0.20763	-0.08614	-0.11456	0.69631	0.29446	0.39081	-0.27730
V, mm/c	-0.39334	-0.31167	-0.03730	-0.43519	0.60847	-0.29888	-0.77029
h, mm	0.25979	-0.05090	-0.26377	0.26111	-0.06161	-0.17481	-0.33893

3.3. Gray Relational Analysis

Gray relational analysis is used to analyze the quality indicators (experiment response characteristics) due to their multi-directionality (different degrees of desirability "more is better" and "less is better"). Table 7 shows the normalized experiment response characteristics obtained using Formulas (1) and (2).

Table 7. Normalized values of the response characteristics of the experiment.

Experiment Number	Yield Stress, $\sigma_{0.2}$, MPa	Tensile Strength, σ_B, MPa	Relative Elongation, δ, %	Microstructure Desirability Rank	Non-Melting Area, mm^2		Pore Diameter, mm
					In the Central Part	Along Edges	
1	1.000	1.000	0.640	0.000	0.864	0.236	0.333
2	0.241	0.000	0.000	0.000	0.992	1.000	0.333
3	0.096	0.349	0.407	0.500	0.996	0.836	0.000
4	0.322	0.767	1.000	0.000	0.999	0.779	0.467
5	0.276	0.392	0.467	0.250	1.000	0.683	0.000
6	0.238	0.570	0.653	0.250	0.959	0.000	0.400
7	0.096	0.193	0.300	1.000	0.000	0.904	0.533
8	0.365	0.606	0.793	0.000	1.000	0.575	0.667
9	0.000	0.034	0.120	1.000	0.268	0.745	1.000

The calculated values of the gray relational coefficient ξ_{ij} obtained using Formula (3) are presented in Table 8. The resulting relational quality scores obtained using Formula (4) are shown in Figure 6. The higher the integrated relational score, the better the result of the experiment, and the closer it is to the ideally normalized value. According to the calculated values, it can be seen that the fourth (0.688) sample has the maximum value of the integral relational assessment. Therefore, the best quality performance is achieved by using a combination of the following optimal scanning parameters: laser power of 250 W and scanning speed of 1200 mm/s, which corresponds to a VED of 34.7 J/mm^3.

Table 8. The values of the gray relational coefficient ξ_{ij}.

Experiment Number	Yield Stress, $\sigma_{0.2}$, MPa	Tensile Strength, σ_B, MPa	Relative Elongation, δ, %	Microstructure Desirability Rank	Non-Melting Area, mm^2		Pore Diameter, mm
					In the Central Part	Along Edges	
1	0.100	0.140	0.280	0.500	0.068	0.382	0.333
2	0.480	0.840	0.600	0.500	0.004	0.000	0.333
3	0.552	0.596	0.397	0.250	0.002	0.082	0.500
4	0.439	0.303	0.100	0.500	0.001	0.110	0.267
5	0.462	0.566	0.367	0.375	0.000	0.159	0.500
6	0.481	0.441	0.273	0.375	0.021	0.500	0.300
7	0.552	0.705	0.450	0.000	0.500	0.048	0.233
8	0.418	0.416	0.203	0.500	0.000	0.212	0.167
9	0.600	0.816	0.540	0.000	0.366	0.127	0.000

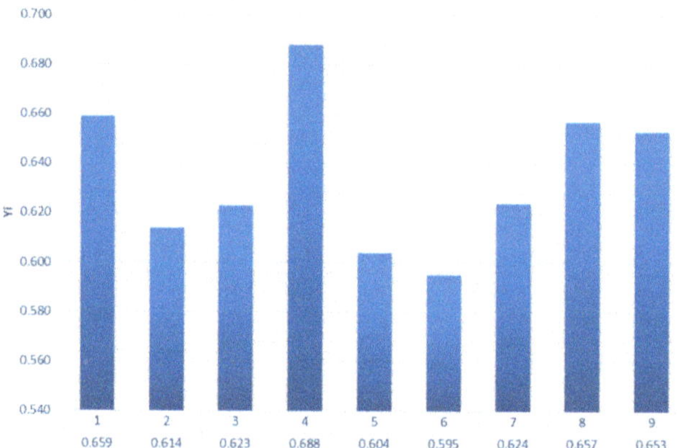

Figure 6. The resulting relational quality scores.

Specimen geometry has to comply with ISO 6892-1:2019 and ASTM E8/E8M-16a standards. The Instron 8802 testing machine was used to evaluate the tensile properties at room temperature, with a displacement rate of a crosshead of 0.0075 mm/s. The uniaxial tensile testing results obtained for the cylindrical samples are summarized in Table 9.

Table 9. Mechanical properties of cylindrical samples.

Experiment Number	Yield Stress, $\sigma_{0.2}$, MPa	Tensile Strength, σ_B, MPa	Relative Elongation, δ, %
1	620.3	795.1	19.5
2	625.3	801.4	21
3	618.1	799.7	20.7
Average values	621.2	798.7	20.4
Casting alloy [37]	703.0	768.0	3.3

4. Conclusions

In this work, gray relational analysis was used to determine the optimal scanning parameters for the Ni-Cr-Al-Ti-based superalloy powder. In order to obtain an overall quality score for multiple quality characteristics of the SLM process, a single one called grey relational grade was used with a simplified optimization procedure. The subsequent analyses and the experiments design technique are carried out to evaluate the optimal parametric combinations. The possibility of the simultaneous optimization of several quality parameters (yield stress, tensile strength, relative elongation, non-melting area, pore diameter, etc.) has been demonstrated.

The analysis of the obtained results made it possible to establish the following:

- The level of tensile strength on the test samples varied from 774.7 to 949.0 MPa, with a relative elongation from 5.2% to 10.2%;
- The highest tensile strength (949 MPa), with an average level of plasticity (8.4%), was obtained on the samples manufactured using technological modes corresponding to a VED of 66.7 J/mm^3 at a laser power of 300 W, a scanning speed of 600 mm/s, and a hatch distance of 0.15 mm;
- The highest ductility (10.2%), with a high level of tensile strength (908.4 MPa), was observed on the samples manufactured using technological modes corresponding to a VED of 34.7 J/mm^3 at a laser power of 250 W, a scanning speed of 120 mm/s, and a hatch distance of 0.12 mm.

The correlation analysis results, as applied to the technological regime influence analysis of the sample growth on the large pore formation, showed that the optimal technological parameters for the series of performed experiments belonged to the scope of the lower VED values and the higher scanning speeds. For a more detailed analysis and selection of the optimal mode for the growing samples, taking into account the multiple quality characteristics, a gray relational analysis was carried out.

As a result of the gray relational analysis, it was found that the optimal values of the mechanical properties, microstructure, and defect sizes were achieved by combining the following technological scanning parameters: laser power of 250 W and scanning speed of 1200 mm/s.

The results of the uniaxial tensile tests of the cylindrical samples made using the optimal scanning parameters made it possible to establish that the mechanical properties of the Ni-Cr-Al-Ti-based heat-resistant alloy synthesized by the SLM technology exceeded the mechanical properties of the same material obtained by the casting technology.

So, the average yield strength of the heat-resistant material obtained by the SLM technology was 621.2 MPa, which was 11,6% less than the standard values for this alloy. The average tensile strength of the heat-resistant material obtained by the SLM technology was 798.7 MPa, which was 3,8% higher than the standard values for this alloy. The average relative elongation of the heat-resistant material obtained by the SLM technology was 20.4%, which was 83.8% higher than the standard values for this alloy.

Author Contributions: Conceptualization, V.V.K. and V.G.S.; methodology, A.I.K.; formal analysis, E.V.Z.; investigation, A.V.A.; writing—original draft preparation, K.D.B.; writing—review and editing, A.Y.K. All authors have read and agreed to the published version of the manuscript.

Funding: The research is partially funded by the Ministry of Science and Higher Education of the Russian Federation as part of the World-class Research Center program: Advanced Digital Technologies (contract No. 075-15-2022-312, dated 20 April 2022).

Institutional Review Board Statement: Not applicable.

Informed Consent Statement: Not applicable.

Data Availability Statement: Not applicable.

Conflicts of Interest: The authors declare no conflict of interest.

References

1. Popovich, A.A.; Sufiiarov VSh Borisov, E.V.; Polozov, I.A.; Masaylo, D.V. Design and manufacturing of tailored microstructure with selective laser melting. *Mater. Phys. Mech.* **2018**, *38*, 1–10.
2. Verma, S.; Kumar, A.; Lin, S.-C.; Jeng, J.-Y. A bio-inspired design strategy for easy powder removal in powder-bed based additive manufactured lattice structure. *Virtual Phys. Prototyp.* **2022**, *17*, 468–488. [CrossRef]

3. Gogolewski, D.; Kozior, T.; Zmarzły, P.; Mathia, T.G. Morphology of Models Manufactured by SLM Technology and the Ti6Al4V Titanium Alloy Designed for Medical Applications. *Materials* **2021**, *14*, 6249. [CrossRef]

4. Verma, S.; Yang, C.-K.; Lin, C.-H.; Jeng, J.Y. Additive manufacturing of lattice structures for high strength mechanical interlocking of metal and resin during injection molding. *Addit. Manuf.* **2022**, *49*, 102463. [CrossRef]

5. Sufiiarov VSh Popovich, A.A.; Borisov, E.V.; Polozov, I.A. Layer thickness influence on the inconel 718 alloy microstructure and properties under selective laser melting. *Tsvetnye Met.* **2017**, *1*, 77–82.

6. Sotov, A.V.; Agapovichev, A.V.; Smelov, V.G.; Kokareva, V.V.; Dmitrieva, M.O.; Melnikov, A.A.; Golanov, S.P.; Anurov, Y.M. Investigation of the IN-738 superalloy microstructure and mechanical properties for the manufacturing of gas turbine engine nozzle guide vane by selective laser melting. *Int. J. Adv. Manuf. Technol.* **2020**, *107*, 2525–2535. [CrossRef]

7. Sanchez, S.; Smith, P.; Xu, Z.; Gaspard, G.; Hyde, C.J.; Wits, W.W.; Ashcroft, I.A.; Chen, H.; Clare, A.T. Powder Bed Fusion of nickel-based superalloys: A review. *Int. J. Mach. Tools Manuf.* **2021**, *165*, 103729. [CrossRef]

8. Zhang, H.; Li, C.; Guo, Q.; Ma, Z.; Huang, Y.; Li, H.; Liu, Y. Hot tensile behavior of cold-rolled Inconel 718 alloy at 650 °C: The role of δ phase. *Mater. Sci. Eng. A* **2018**, *722*, 136–146. [CrossRef]

9. Tomus, D.; Rometsch, P.A.; Heilmaier, M.; Wu, X. Effect of minor alloying elements on crack-formation characteristics of Hastelloy-X manufactured by selective laser melting. *Addit. Manuf.* **2017**, *16*, 65–72. [CrossRef]

10. Haafkens, M.H.; Matthey, G.H. A New Approach to Weldability of Nickel-Base As-Cast and Powder Metallurgy Superalloys. *Weld J.* **1982**, *61*, 25.

11. Marchese, G.; Parizia, S.; Saboori, A.; Manfredi, D.; Lombardi, M.; Fino, P.; Ugues, D.; Biamino, S. The Influence of the Process Parameters on the Densification and Microstructure Development of Laser Powder Bed Fused Inconel 939. *Metals* **2020**, *10*, 882. [CrossRef]

12. Sufiyarov, V.S.; Borisov, E.V.; Polozov, I.A.; Masailo, D.V. Control of structure formation in selective laser melting process. *Tsvetnye Met.* **2018**, *7*, 68–74. [CrossRef]

13. Hu, Y.; Kang, W.; Zhang, H.; Chu, C.; Wang, L.; Hu, Y.; Ding, Y.; Zhang, D. Hot corrosion behavior of IN738LC alloy formed by selective laser melting. *Corros. Sci.* **2022**, *198*, 110154. [CrossRef]

14. Li, J.; Hu, J.; Cao, L.; Wang, S.; Liu, H.; Zhou, Q. Multi-objective process parameters optimization of SLM using the ensemble of metamodels. *J. Manuf. Process.* **2021**, *68A*, 198–209.

15. Ahmed, I.; Barsoum, G.; Haidemenopoulos, R.K. Abu Al-Rub, Process parameter selection and optimization of laser powder bed fusion for 316L stainless steel: A review. *J. Manuf. Process.* **2022**, *75*, 415–434. [CrossRef]

16. Kamath, C.; El-Dasher, B.; Gallegos, G.F.; King, W.E.; Sisto, A. Density of additively-manufactured, 316L SS parts using laser powder-bed fusion at powers up to 400 W. *Int. J. Adv. Manuf. Technol.* **2014**, *74*, 65–78. [CrossRef]

17. Brandt, M. *Laser Additive Manufacturing: Materials, Design, Technologies, and Applications*; Woodhead Publishing: Sawston, UK, 2016.

18. Jain, P.K.; Pandey, P.M.; Rao, P.V.M. Effect of delay time on part strength in selective laser sintering. *Int. J. Adv. Manuf. Technol.* **2009**, *43*, 117–126. [CrossRef]

19. Chlebus, E.; Kuźnicka, B.; Kurzynowski, T.; Dybała, B. Microstructure and mechanical behaviour of Ti-6Al-7Nb alloy produced by selective laser melting. *Mater. Charact.* **2011**, *62*, 488–495. [CrossRef]

20. Yu, Z.; Xu, Z.; Guo, Y.; Xin, R.; Liu, R.; Jiang, C.; Li, L.; Zhang, Z.; Ren, L. Study on properties of SLM-NiTi shape memory alloy under the same energy density. *J. Mater. Res. Technol.* **2021**, *13*, 241–250. [CrossRef]

21. Heeling, T.; Gusarov, A.; Cloots, M.; Wegener, K. Melt pool simulation for the evaluation of process parameters in selective laser melting. *Addit. Manuf.* **2017**, *14*, 116–125. [CrossRef]

22. Agapovichev, A.V.; Khaimovich, A.I.; Kokareva, V.V.; Smelov, V.G. Determining Rational Technological Parameters for Selective Laser Melting of AlSi10Mg Aluminum Alloy Powder. *Inorg. Mater. Appl. Res* **2022**, *13*, 543–548. [CrossRef]

23. Gong, H.; Rafi, K.; Gu, H.; Starr, T.; Stucker, B. Analysis of defect generation in Ti–6Al–4V parts made using powder bed fusion additive manufacturing processes. *Addit. Manuf.* **2014**, *1–4*, 87–98. [CrossRef]

24. Vrancken, B. Study of Residual Stresses in Selective Laser Melting. Ph.D. Thesis, KU Leuven, Leuven, Belgium, 2016.

25. Montgomery, C.; Beuth, J.; Webler, B. Melt pool geometry and microstructure of Ti6Al4V with B additions processed by selective laser melting additive manufacturing. *Mater. Des.* **2019**, *183*, 108126.

26. Sohrabpoor, H.; Negi, S.; Shaiesteh, H.; Inamul, A.; Brabazon, D. Optimizing selective laser sintering process by grey relational analysis and soft computing techniques. *Optik* **2018**, *174*, 185–194. [CrossRef]

27. Acherjee, B.; Kuar, A.S.; Mitra, S.; Misra, D. Application of grey-based Taguchi method for simultaneous optimization of multiple quality characteristics in laser transmission welding process of thermoplastics. *Int. J. Adv. Manuf. Technol.* **2011**, *56*, 995–1006. [CrossRef]

28. Khaimovich, A.; Erisov, Y.; Smelov, V.; Agapovichev, A.; Petrov, I.; Razhivin, V.; Bobrovskij, I.; Kokareva, V.; Kuzin, A. Interface Quality Indices of Al–10Si–Mg Aluminum Alloy and Cr18–Ni10–Ti Stainless-Steel Bimetal Fabricated via Selective Laser Melting. *Metals* **2021**, *11*, 172. [CrossRef]

29. Balasubramanian, S.; Ganapathy, S. Grey Relational Analysis to determine optimum process parameters for Wire Electro Discharge Machining (WEDM). *Int. J. Eng. Sci. Technol.* **2011**, *3*, 95–101.

30. Deng, J. Introduction to grey theory. *J. Grey Syst.* **1989**, *1*, 1–24.

31. Luca, A.D.; Kenel, C.; Griffiths, S.; Joglekar, S.S.; Leinenbach, C.; Dunand, D.C. Microstructure and defects in a Ni-Cr-Al-Ti γ/γ' model superalloy processed by laser powder bed fusion. *Mater. Des.* **2021**, *201*, 109531. [CrossRef]

32. Dursun, G.; Orhangul, A.; Urkmez, A.; Akbulut, G. Understanding the Parameter Effects on Densification and Single Track Formation of Laser Powder Bed Fusion Inconel 939. *Procedia CIRP* **2022**, *108*, 258–263. [CrossRef]

33. Chua, C.K.; Leong, K.F. *3D Printing Additive Manufacturing: Principles Applications*, 4th ed.; World Scientific Publishing: Singapore, 2015.

34. Dowden, J. *The Theory of Laser Materials Processing*; Springer: Berlin/Heidelberg, Germany, 2009.

35. King, W.E.; Barth, H.D.; Castillo, V.M.; Gallegos, G.F.; Gibbs, J.W.; Hahn, D.E.; Kamath, C.; Rubenchik, A.M. Observation of keyhole-mode laser melting in laser powder-bed fusion additive manufacturing. *J. Mater. Process. Technol.* **2014**, *214*, 2915–2925. [CrossRef]

36. Eagar, T.; Tsai, N. Temperature Fields Produced by Traveling Distributed Heat Sources. *Weld. J.* **1983**, *62*, 346–355.

37. Jahangiri, M.R.; Abedini, M. Effect of long time service exposure on microstructure and mechanical properties of gas turbine vanes made of IN939 alloy. *Mater. Des.* **2014**, *64*, 588–600. [CrossRef]

 *materials*

Article

Comparison of Aerosol Pt, Au and Ag Nanoparticles Agglomerates Laser Sintering

Kirill Khabarov *, Messan Nouraldeen, Sergei Tikhonov, Anna Lizunova, Olesya Seraya, Emiliia Filalova and Victor Ivanov *

Moscow Institute of Physics and Technology, National Research University, 141701 Dolgoprudny, Russia; messannouraldeen@phystech.edu (M.N.); sergei.s.tikhonov@phystech.edu (S.T.); Lizunova.aa@mipt.ru (A.L.); seraia.ov@phystech.edu (O.S.); filalova.em@phystech.edu (E.F.)
* Correspondence: kirill.khabarov@phystech.edu (K.K.); ivanov.vv@mipt.ru (V.I.)

Abstract: In this paper, we investigated the interaction of nanosecond pulsed-periodic infrared (IR) laser radiation at a 50 and 500 Hz repetition rate with aerosol platinum (Pt) and silver (Ag) nanoparticles agglomerates obtained in a spark discharge. Results showed the complete transformation of Pt dendrite-like agglomerates with sizes of 300 nm into individual spherical nanoparticles directly in a gas flow under 1053 nm laser pulses with energy density 3.5 mJ/cm^2. Notably, the critical energy density required for this process depended on the size distribution and extinction of agglomerates nanoparticles. Based on the extinction cross-section spectra results, Ag nanoparticles exhibit a weaker extinction in the IR region in contrast to Pt, so they were not completely modified even under the pulses with energy density up to 12.7 mJ/cm^2. The obtained results for Ag and Pt laser sintering were compared with corresponding modification of gold (Au) nanoparticles studied in our previous work. Here we considered the sintering mechanisms for Ag, Pt and Au nanoparticles agglomerates in the aerosol phase and proposed the model of their laser sintering based on one-stage for Pt agglomerates and two-stage shrinkage processes for Au and Ag agglomerates.

Keywords: aerosol nanoparticles; platinum; gold; silver; spark discharge; pulsed laser radiation; IR; shrinkage; plasmon resonance; extinction

Citation: Khabarov, K.; Nouraldeen, M.; Tikhonov, S.; Lizunova, A.; Seraya, O.; Filalova, E.; Ivanov, V. Comparison of Aerosol Pt, Au and Ag Nanoparticles Agglomerates Laser Sintering. *Materials* **2022**, *15*, 227. https://doi.org/10.3390/ma15010227

Academic Editor: Alina A. Manshina

Received: 10 December 2021
Accepted: 27 December 2021
Published: 29 December 2021

Publisher's Note: MDPI stays neutral with regard to jurisdictional claims in published maps and institutional affiliations.

1. Introduction

The optical and electrical properties of metal nanoparticles (NPs) are interrelated and largely determined by the electronic configuration of the material and their sizes [1]. For example, in NPs consisting of hundreds–thousands of atoms, electronic states become discrete due to the electron mean free path (MFP) limiting at the surface and grain boundaries [2,3]. For this reason, the spectral characteristics and some parameters as electrical conductivity and magnetic susceptibility of such NPs change significantly and exhibit quantum-dimensional effects [4]. Significant nonlinearity in the parameters of nanoscale metals is also observed for larger dimensions of electrons' spatial localization [5]. Simultaneously with the nanoobjects' size parameters, the mobility of electrons is determined by the material electronic configuration affecting the intrinsic MFP of charge carriers. For example, the electrons MFP for platinum group metals (Pt, Ir, Pd, Ru) turn out to be shorter than for noble metals (Au, Ag, Cu). As a result, such metals may show lower resistivity in nanoscale due to less scattering at the nanoobjects surface and grain boundaries [5]. In this regard, the study of the spectral characteristics of isolated NPs, severely limited in size in any of the directions, is an urgent task [6,7].

Recently, NPs properties have been studied by spectrophotometric approaches, e.g., X-ray photoelectron spectroscopy (XPS) [8], ultraviolet photoelectron spectroscopy (UPS) [9] and others. However, the results of the presented methods' research significantly depends on the environment inevitably interacting with NPs, and this plays a crucial role. For instance, the electronic structure of NPs changes greatly in colloidal solutions, which leads to significant shifts in optical resonances and difficulties in obtaining accurate results [10]. This creates the need to develop approaches for the study of NPs in a gas phase. These approaches are promising for obtaining high-precision results that are in good agreement with the theory. In our previous work [11], we have already proposed such an approach and studied optical spectra and processes of laser radiation effect on Au NPs agglomerates expressed in their sintering directly in a gas flow. However, the comparison of this metal with Pt and Ag is also of great interest and defines the novelty of the work considering the sintering mechanisms of nanoparticles right in the aerosol phase. These materials are of fundamental importance for a variety of catalytic [12] and plasmonic [13] applications such as photocatalysis [14–16], synthesis of reactive oxygen species [17,18], imaging [13,19], detection of chemical and biological substances [20], and others.

In this paper, we experimentally studied the extinction cross-section spectra and sintering processes by pulsed nanosecond laser radiation with a wavelength of 1053 nm for aerosol agglomerates of NPs synthesized in a spark discharge (SD). Agglomerate sintering is caused by the diffusion flow of matter in a direction that provides minimal surface energy, which can be achieved with a spherical shape [21]. To do this, we used a previously developed laser modification cell [11], which combined the studied aerosol flow with the optical axis of impacting radiation. In the experiments, the NPs agglomerates were spatially separated—their interactions with each other and with the experimental setup were minimized. The study was done for Pt and Ag NPs materials significantly differ in optical and electrical properties. Here we measured the NPs agglomerates' average size and concentration online by their differential electrical mobility during their laser modification, depending on the energy density of laser radiation pulses. Along with laser exposure, we studied the extinction cross-section of these NPs agglomerates in a wide spectral range by a spectrophotometer. The obtained results for Pt and Ag were compared to results newly obtained for Au NPs (TEM images, NPs distribution and extinction cross-section spectra) and taken from our previous work (size-energy density dependencies) [11].

2. Materials and Methods

Primary NPs were synthesized in the SD during the electrical erosion of the electrodes in an atmosphere of high purity argon (99.9999%) at an excess gas pressure of 0.6 atm. The operation principles of a NPs SD generator with all the explanations and parameters of the electrical scheme used are described in detail in [22]. The electrodes used in the work were made of Pt, Au and Ag with a purity of 99.9999% and had a shape of hollow cylinders with an outer diameter of 8 mm (Plaurum Group, Verhnyaya Pyshma, Russia). In the case of Pt and Au, the thickness of the cylinder wall was 1 mm, and, in the case of Ag, 3 mm. The high purity of the metal and gas medium together with the vacuum tightness of the gas path and synthesis chamber ensured the maximum possible purity of the studied NPs during the experiments.

The study of the optical radiation interaction with aerosol NPs agglomerates was carried out in the cell discussed in detail in [11]. In the experiments, we used a pulsed laser with the wavelength of 1053 nm (TECH-1053, "Laser-export" Co. Ltd., Moscow, Russia) with a pulse duration of about 40 ns and controlled pulse repetition rates in the range of 10 Hz–10 kHz. Here we conducted the experiments at the pulse repetition rates of 50 and 500 Hz with equal pulse energy range up to 900 µJ.

Schematically, the experimental setup is shown in Figure 1.

In all the experiments, the laser was initially set up at the position of the light source in front of the cell so that the radiation was directed opposite to the aerosol flow. The gas flow rate used (Q = 50 mL/min) determined the velocity of NPs inside the cell υ = 117.9 mm/s. The diameter of the laser beam in the cell was 3 mm. The study of the laser radiation interaction with NPs was done under the control of the average radiation power by the thermal power meter PD300-3W-V1 (Ophir, North Logan, UT, USA). The energy density of a single laser pulse was calculated based on the measured average radiation power and known pulse repetition rate, provided that the laser generates the same pulses [11].

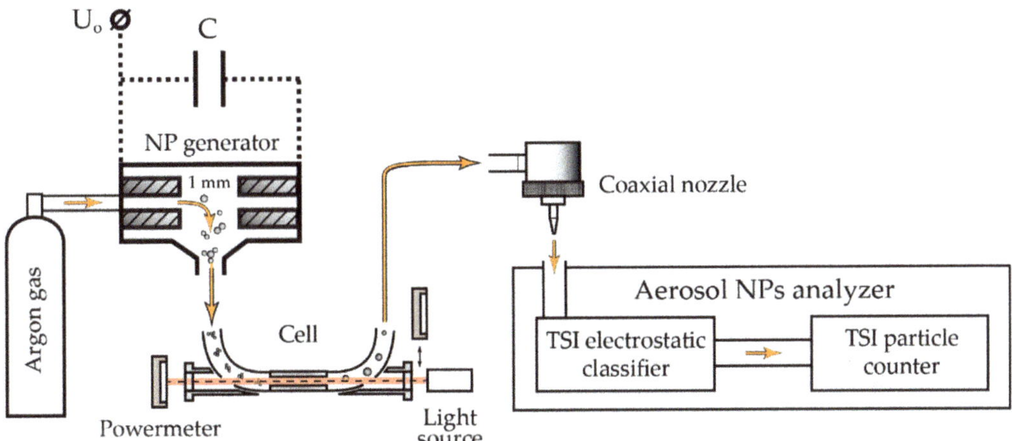

Figure 1. The scheme of the experimental setup.

The optical spectra of aerosol NPs agglomerates were studied in the wavelength range of 300–1000 nm by the spectrophotometer HR4000CG-UV-NIR (Ocean Insight, Orlando, FL, USA) with a white light source (halogen lamp), equipped with an optical fiber, that was installed in front of the cell. For this purpose, for each material, we did successive measurements of the halogen lamp spectrum through the cell, initially filled only with the gas, and then with the aerosol. The measurements were done with the fixed geometry of the experiment and stable position of the optical elements. Final spectra of aerosol NPs were obtained by normalizing spectra for the cell filled with the aerosol on spectra for the cell filled with the gas only. This procedure excluded the influence of possible reflections of radiation inside the cell on the results.

Statistical size distributions of aerosol NPs, as well as their total concentrations in the gas flow, were controlled online after the particles left the cell through a focusing nozzle by the direct measurements of the aerosol NPs analyzer SMPS 3936 (TSI Inc., Shoreview, MN, USA) based on their differential electrical mobility. The statistical size distributions of aerosol NPs obtained in these measurements were described by a log-normal distribution. To construct the dependences of the agglomerates size on the energy density of laser pulses, we approximated the size distributions with a set of log-normal peaks, controlling $R^2 > 0.95$, and measured the maxima of the highest ones (Figure 2). The focusing nozzle with an outlet diameter of 300 µm simultaneously delivered NPs to the analyzer and stabilized the parameters of pressure and flow of the carrier gas. It is worth mentioning that possible losses of NPs, due to their deposition on walls of a pipe between the cell and the analyzer with a length of about 200 mm, caused by the diffusion processes, weakly affecting the concentration measurements, are less than 3% and are within the error of the analyzer [11].

Additional studies of shapes and sizes of NPs were done based on images obtained by the transmission electron microscope (TEM) JEM-2100 (JEOL, Ltd., Tokyo, Japan). To do this, NPs were collected on TEM grids installed at the focusing nozzle outlet right after the cell. According to the set of obtained TEM images, we plotted size distributions and measured modal sizes for primary synthesized NPs in the agglomerate composition and for spherical NPs sintered by laser radiation for the three metals by processing 600 pieces on average, approximating them with spheres.

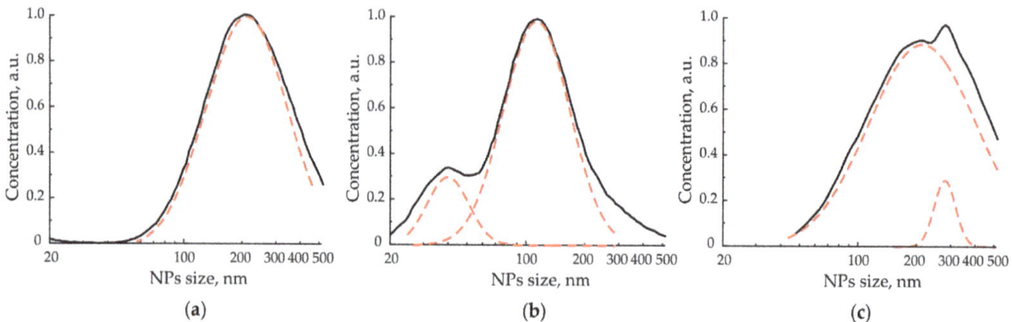

Figure 2. Examples of size distributions for initial (**a**) and laser modified (**b**,**c**) NPs agglomerates (black curves) with their log-normal multiple peak approximation (dashed red curves). Here, the NPs size axis is provided in the logarithmic scale.

3. Results

3.1. NPs Spectral Characteristics

During the transportation in the aerosol flow, primary NPs synthesized in the SD experience many collisions due to their Brownian motion and assemble into dendrite-like agglomerates of arbitrary shape characterized by a log-normal size distribution with a modal size of 150–300 nm (Figure 3). The agglomerates size depends on the velocity parameters of the aerosol determining the NPs flow time, initial NPs concentration, and the sizes of primary NPs for each material. In our study, the modal sizes of Pt, Au and Ag agglomerates measured by the aerosol spectrometer were 300, 280 and 167 nm, respectively. At the same time, the modal sizes of primary NPs were analyzed from a series of TEM images and were found equal to 3.5, 7.5 and 17.5 nm for Pt, Au and Ag, respectively. As a result, Au and Ag NPs agglomerates had larger necks connecting primary NPs.

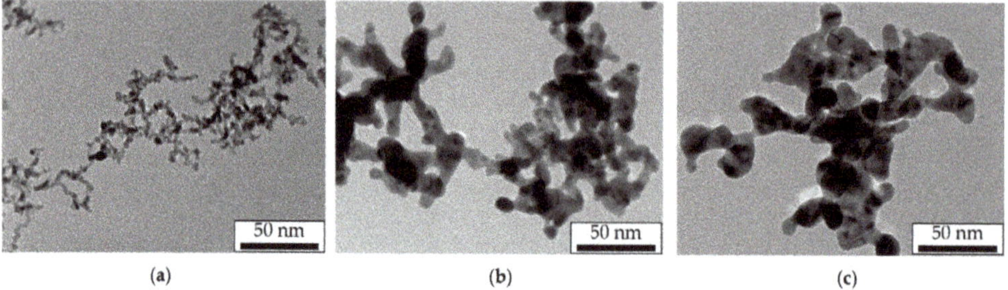

Figure 3. TEM images of Pt (**a**), Au (**b**) and Ag (**c**) NPs agglomerates.

Aerosols of initial agglomerates were studied by the spectrophotometer for all the three materials. Based on the obtained transmission spectra for the initial agglomerates, we determined the extinction cross-section spectra of one agglomerate (Figure 4), averaged over the aerosol flow, calculated according to the Beer–Lambert law [23].

We also determined the extinction cross-section of initial agglomerates at the wavelength of 1053 nm radiating them by laser pulses with the energy density 3 mJ/cm^2 and pulse repetition rate of 500 Hz at the gas flow rate 200 mL/min. With these parameters, the agglomerates were still unmodified. Extinction cross-sections equaled 0.105, 0.072 and 0.005 μm^2 for Pt, Au and Ag, respectively, which is comparable to the literature data for nanostars studied in colloidal solutions [24]. Neglecting the fact that the agglomerates size could slightly change due to the change in the gas flow, in the further discussion, we will use these cross-sections. With comparable sizes of NPs agglomerates for Pt and Au, Pt agglomerates had the largest extinction cross-section, and the extinction cross-section of Ag NPs turned out to be the smallest. The result does not change even if we take into account the less than 2 times smaller modal size of Ag NPs agglomerates, since their extinction cross-section is 21 and 14 times smaller than for Pt and Au at the laser wavelength, respectively.

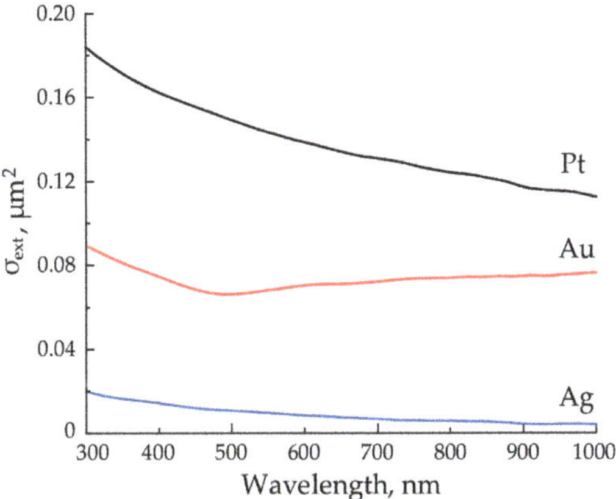

Figure 4. Extinction cross-section spectra of a single NP averaged over the flow of initial agglomerates for Pt, Au and Ag.

3.2. NPs Agglomerates Laser Modification

We studied the effect of nanosecond pulse-periodic laser radiation with the wavelength of 1053 nm on the size of sintered NPs agglomerates depending on the pulse energy density at the repetition rates of 50 and 500 Hz. The sizes were measured in the gas flow by the NPs electrical mobility. Similarly to [11], we will call the sintering of NPs in the aerosol flow complete if, as a result of their modification, only spherical particles remain, which does not change with a further increase in the pulse energy density. Otherwise, we will consider the sintering process incomplete.

The morphology of NPs modified by laser radiation at the maximum pulse energy density and pulse repetition rate of 500 Hz is shown in the TEM images presented in Figure 5. Under the close sintering conditions, the Pt and Au agglomerates were completely sintered, which is confirmed by the large spherical NPs in Figure 5a,b. However, for Ag, one can see dendritic structures in the TEM images (Figure 5c), which characterize the sintering as incomplete.

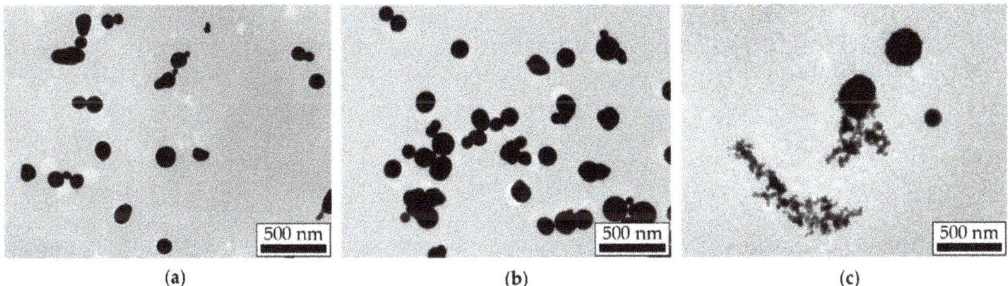

Figure 5. TEM images of Pt (**a**), Au (**b**) and Ag (**c**) NPs agglomerates, modified by laser radiation with the wavelength of 1053 nm at the maximum pulse energy density and pulse repetition rate of 500 Hz for the aerosol flow rate 50 mL/min.

During the pulsed laser sintering of aerosol NPs agglomerates, we measured the sizes of sintered NPs in the stream by their electrical mobility, depending on the energy density of the incident radiation pulse (Figure 6a–c). In this figure, the black dashed line corresponds to the distribution modal size of the initial agglomerates, determined by their electrical mobility. The red dashed line corresponds to the modal size of sintered agglomerates determined by statistical processing of the part of sintered spherical NPs on a series of TEM images for each of the three materials. As for Ag NPs, the measured modal size may slightly differ from the actual modal size of completely modified NPs and is shown on the graph for result comparison. This is shown by the small red peak in Figure 6f with a maximum of around 40 nm.

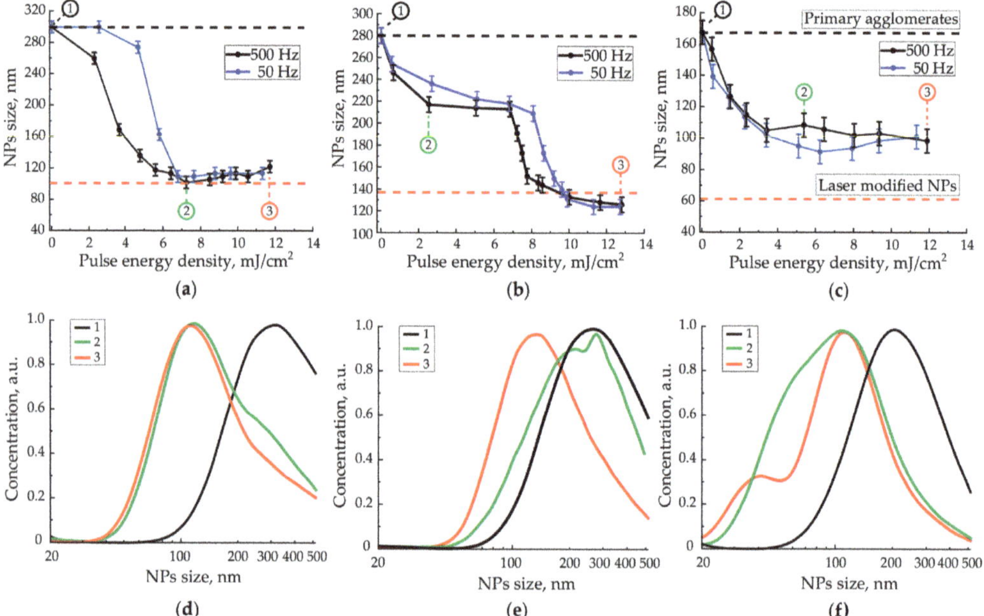

Figure 6. Dependencies of Pt (**a**), Au (**b**) and Ag (**c**) NPs size at the output of the focusing nozzle right after the cell on the pulse energy density of laser radiation with the wavelength of 1053 nm at the pulse repetition rate of 500 Hz and the aerosol flow rate of 50 mL/min. The shrinkage curve data for Au is from [11]. Size distributions of initial and laser modified Pt (**d**), Au (**e**) and Ag (**f**) NPs agglomerates taken from 1, 2 and 3 measurements presented in (**a–c**).

Experimental dependencies of the NPs agglomerates' size for the three metals at the exit from the cell (Figure 6) demonstrate a decrease in the size as the energy density of laser pulses increases. This process should be called NPs shrinkage as the result of laser sintering in analogy with the shrinkage of macroscopic powder green bodies under the energy exposure to them [25].

On the presented curves corresponding to Pt and Au, one may observe an S-shaped behavior with an initially slow decrease in size to a certain threshold value of the pulse energy density, then with a rapid decrease in size in a narrow energy density range and, finally, a slow decrease with an output to a constant value. However, on similar dependencies for Ag NPs, the rapid size reduction in a narrow energy density range was not achieved, and sintering turned out to be incomplete. This is also confirmed by the TEM images of laser modified Ag NPs (Figure 5c).

4. Discussion

The shrinkage of Pt NPs agglomerates caused by laser radiation, shown in Figure 6a, is characterized by a slow decrease in size at the pulse energy density values below 2.5

and 4.5 mJ/cm^2 at the pulse repetition rates of 500 and 50 Hz, respectively. With a further increase in the pulse energy density, a steep shrinkage interval is observed, reaching a horizontal section with a final NPs size of about 100 nm at the pulse energy densities of 6.0 and 6.7 mJ/cm^2 and pulse repetition rates of 500 and 50 Hz, respectively. According to aerosol in-flow particle size analysis, the unimodal agglomerate distribution was observed for three main stages of the Pt shrinkage process: 1st at the beginning of the experiments when measuring initial agglomerates, 2nd after the end of the first step of sintering, when the shrinkage curve flattened for the first time, 3rd at the maximum pulse energy density observed in our experiments. Comparing Pt NPs size distributions, we discover a 200 nm-shift of the unimodal peak to the low-size range with the growth of the energy density (Figure 6d). One may find in thermal sintering of powder green bodies prepared from identical particles with a unimodal size distribution a similar one-stage shrinkage process with one section of steep size change. In this case, particle sintering requires close energies [25]. Based on the presented analogy and the presence of only one type of agglomerates, we can assume that the absorption characteristics of radiation at the wavelength of 1053 nm are close for all Pt NPs agglomerates.

It is worth noticing that Pt NPs agglomerates require the lowest pulse energy density for complete laser sintering among other materials. This can be explained by a comparison of optical properties for metal NPs. Small Pt NPs have an approximately constant high-level absorption 0.4 in red and IR range starting from around 500 nm wavelength [26]. Close results were received in a simulation of optical radiation absorption by thin Pt films with a thickness of about 10 nm on glass surfaces [27]. In the above-noted work, the thin films were characterized by a high and stable absorption of about 0.4 over a wavelength range of 250–900 nm. Hereafter, because of the lack of experimental and theoretical research on the absorption of IR radiation by NPs, and taking into account the noted data about Pt films and particles, we present the literature data about electrical and optical properties of thin films and consider them close to NPs ones.

Experimental and theoretical studies of the electrical resistivity of polycrystalline Pt films [5] showed that, with a decrease in the thickness of Pt films from 40 to 3 nm, their electrical resistivity increased by less than 2 times. Moreover, Pt thin films showed the lowest resistivity for films with a thickness below 5 nm in comparison with other etalon conductive materials such as copper. This effect was explained by a weaker thickness dependence of both surface and grain boundary electron scattering in Pt and connected with the efficiency of absorption.

In contrast to Pt, the interaction process of IR laser radiation with Au agglomerates was characterized by a clearly expressed two-stage shrinkage with two areas of steep size change (Figure 6b) from our work [11]). The first section of steep shrinkage starts at the minimum values of the energy density of laser pulses and ends at about 2.5 and 5.0 mJ/cm^2 at the pulse repetition rates of 500 and 50 Hz, respectively. The second steep shrinkage phase is observed in the ranges of pulse energy densities 7.0–8.0 mJ/cm^2 and 8.3–10.0 mJ/cm^2 for the pulse repetition rates 500 and 50 Hz, respectively. The shift of a steep section of the shrinkage curves towards higher values of the energy density for both Au and Pt caused by the reduction of the pulse repetition rate is associated with a discrete-stepwise process of NPs shrinkage under the influence of laser pulses. In particular, this happens because of the fewer number of interacting pulses requiring a higher energy density to modify NPs [11]. Two-stage shrinkage processes were also observed during thermal sintering of powder green bodies prepared from particles with a two-modal size distribution [28], as the particles with a smaller characteristic size require less energy for sintering than large particles.

During the analysis of the evolution of Au and Ag NPs size distribution, we found that initially unimodal agglomerates distribution at the end of the first sintering step splits into a bimodal one (Figure 6e,f). This phenomenon is relative to the formation of two agglomerates fractions with different sizes, and we suggested that some of the initial agglomerates have been completely sintered at this step. The difference in the absorption efficiency of IR radiation between large agglomerates and metal nanospheres explains the observed effect. In [29], the authors have shown that, during the agglomeration process, the plasmon peak around 526 nm of separate Au NPs decreased rapidly. Simultaneously, agglomerates exhibited several peaks, red-shifting, merging and overlapping while the agglomerate size grew. This resulted in the raising and flattening of the absorption in the visible and IR spectrum range. In addition, the growth of the number of NPs in agglomerates composition from 2 to 15 led to a two times increase of IR absorbance, whereas changing the size of agglomerate from 15 to 300 particles resulted in more than three times enlargement of IR absorbance. In addition [30], the absorbance of branched Ag nanostars was characterized by a higher IR absorption comparably to nanospheres with the diameter of 20 nm and differed for stars with various branches lengths [31]. Thus, we suggest that our aerosol ensemble of initial Au and Ag NPs agglomerates consists of two types of agglomerates with different IR absorbance efficiency at the wavelength of 1053 nm that is associated with their different fractal structure.

This assumption is also confirmed by previous studies of optical properties of thin polycrystalline films. In particular, in [32], the optical losses of radiation with wavelengths of 250–2000 nm increased significantly with the reduction of the film thickness from 80 to 20 nm and of crystallites size in their composition. The two-stage shrinkage of Au NPs agglomerates observed in our results may be caused by a steep dependence of optical absorption on the agglomerate shape and size of NPs in the agglomerate that can lead to a more efficient laser heating of smaller or branchy particles. Moreover, Au NPs agglomerates were sintered completely at higher radiation energy densities in comparison with Pt due to a smaller level of optical radiation absorption in the near-IR spectrum region. Earlier, in [27], the authors registered the maximum relative optical absorption for Au films with a thickness of about 10 nm at the level of 0.07 in this spectral range. This value is close to a 0.03 absorption level, which was measured for 20–30 nm Au nanospheres in an IR spectrum [33].

Although Ag also belongs to the group of noble metals as well as Au, and the properties of their bulk materials are similar, in our experiments, we found a significant difference in the shrinkage behavior of Ag NPs agglomerates under the influence of pulsed IR laser radiation. Only one section of steep shrinkage was observed in the dependencies (Figure 6c), which started at the minimum pulse energy density values and ended at about 4.0 mJ/cm^2 for both 500 and 50 Hz laser pulse repetition rates. Moreover, a deep size shrinkage of about 1.7 times in this energy density range still was still far from the red dashed line corresponding to the quasi-completely sintered NPs. The incomplete sintering was additionally confirmed by the TEM images of NPs modified with the maximum pulse energy density (Figure 5c). With the further increase in the energy density to 12.0 mJ/cm^2, agglomerate shrinkage practically did not occur.

Analyzing Au and Ag agglomerates distributions in different sintering stages, one initial peak divides into two peaks with close maximums, which are presumably connected to different types of agglomerates (Figure 6e,f). Moreover, the magnitudes of the peaks continuously changed. While the energy density increased, the concentration of completely sintered agglomerates with lower sizes increased and consequently the number of large agglomerates, which have not been sintered yet, decreased. This led to the segregation and growth of a peak for already sintered agglomerates. After laser interaction with Pt and Au agglomerates, we observed the complete sintering of materials that accompanied lognormal size distribution function conservation. However, in intermediate stages, after the sintering of first type of agglomerates, the lognormal unimodal distribution could be transformed to a multimodal one due to the non-equilibrium process of the energy transfer and the difference in IR laser radiation interaction with agglomerates of various shape and crystallites of different sizes. Thus, within the framework of a single-stage shrinkage of Pt, the sintering was done with the conservation of a lognormal distribution, and within the framework of a two-stage shrinkage of Au and Ag, in an intermediate stage, we observed a change in the distribution function and incompleteness of sintering.

Taking into account the bimodal size distribution of Ag NPs at the energy density of 12 mJ/cm^2 compared to Au, we may conclude that, by applying a higher energy density of laser pulses with the wavelength of 1053 nm, the complete sintering of agglomerates can be achieved. Unfortunately, it could not be realized with our laser, so the second section of steep shrinkage of agglomerates, leading to complete sintering of all Ag NPs, is expected. A large difference in the pulse energy density (more than 8.0 mJ/cm^2) to the expected second shrinkage step may say that Ag nanocrystals and agglomerates perform a strong dependence of optical radiation absorption on their size with a small absolute absorption value. Indeed, similar data were previously obtained in several papers. Thus, in [34], it was shown that the Ag film with the thickness below 8 nm displayed a non-metallic behavior in the wavelength range from 600 to 1000 nm. When the thickness was larger than 12 nm, the extinction coefficient was stable and became closer to those of bulk Ag. Moreover, when the film thickness changed from 6.0 to 4.7 nm, the extinction coefficient decreased by more than 10 times. In addition, in [27], for Ag films with a thickness of about 10 nm optimized for maximum absorption, a relative optical absorption in the near-IR range below 0.01 was found, which was about seven times less than the absorption of Au under similar conditions.

The sintering results of aerosol Pt, Au and Ag NPs agglomerates by nanosecond laser pulses of 1053 nm together with the results of other authors are in good agreement with that determined in our experiments' extinction cross sections for such agglomerates, presented in Figure 4. The extinction values for Pt, Au and Ag NPs agglomerates correlate well with the laser radiation complete sintering efficiency. Moreover, at the laser wavelength, the extinction of Au agglomerates is 1.5 times less in comparison with Pt, and the extinction of Ag agglomerates is 14 times less in comparison with Au.

5. Conclusions

In this paper, we have studied the laser sintering mechanisms of Pt and Ag NPs agglomerates obtained in the SD. The sintering was done by the nanosecond pulsed-periodic laser radiation with the wavelength of 1053 nm, pulse energy density range up to 12.7 mJ/cm^2 and repetition rates of 50 and 500 Hz directly in the gas flow. We obtained the shrinkage curves for NPs agglomerates of the two metals (Pt and Ag) that represent the dependence of the modified NPs agglomerates size on the laser pulse energy density and compared them with the one for Au NPs, discussed in our previous work.

In our experiments, we registered the complete modification of NPs agglomerates into separate spherical NPs only for Pt and Au at energy densities above 3.5 and 7.0 mJ/cm^2 with shrinkage in size by 2.9 and 2.3 times, respectively. Moreover, the shrinkage for Au NPs demonstrated a two-stage laser modification process, whereas, for Pt NPs, this process passed in one stage. Thus, within the framework of a single-stage Pt shrinkage, sintering proceeded with the conservation of a unimodal lognormal agglomerate size distribution with a mode shift only. In contrast, according to a proposed two-stage shrinkage model for Au and Ag, in an intermediate stage, we observed a transformation of the distribution into a bimodal one and incompleteness of sintering for Ag even under the pulses with energy density up to 12.7 mJ/cm^2. We associated the two-stage process for Au and Ag with the existence of two types of agglomerates characterized by the different IR absorption efficiency connected with the different fractal structure of agglomerates and crystallite size within the agglomerates.

It was found that the energy required for complete laser modification correlates well with the extinction of NPs increasing for materials in the Pt, Au, Ag order. Thus, according to the experimental results, Ag NPs with an extinction coefficient 14 times less than one for Au NPs at 1053 nm have not been completely modified. At the same time, we detected only partial transformation of agglomerates into spheres with the relative shrinkage to the size of 100 nm approximately equal to 1.7 at the energy density of 12.0 mJ/cm^2.

Author Contributions: Conceptualization, K.K., V.I. and A.L.; methodology, K.K.; validation, K.K. and V.I.; formal analysis, K.K. and S.T.; investigation, K.K., M.N., O.S., E.F. and S.T.; resources, V.I. and A.L.; data curation, O.S., E.F. and S.T.; writing—original draft preparation, K.K., A.L. and V.I.; writing—review and editing, K.K., A.L. and V.I.; visualization, K.K. and S.T.; supervision, V.I.; project administration, V.I.; funding acquisition, V.I. All authors have read and agreed to the published version of the manuscript.

Funding: This research was funded by the Ministry of Science and Higher Education of the Russian Federation, Contract No. 13.2251.21.0025, Grant No. 075-15-2021-960.

Conflicts of Interest: The authors declare no conflict of interest.

References

1. Rao, C.N.R.; Kulkarni, G.U.; Thomas, P.J.; Edwards, P.P. Metal Nanoparticles and Their Assemblies. *Chem. Soc. Rev.* **2000**, *29*, 27–35. [CrossRef]
2. Kano, S.; Azuma, Y.; Maeda, K.; Tanaka, D.; Sakamoto, M.; Teranishi, T.; Smith, L.W.; Smith, C.G.; Majima, Y. Ideal Discrete Energy Levels in Synthesized Au Nanoparticles for Chemically Assembled Single-Electron Transistors. *Am. Chem. Soc. Nano* **2012**, *6*, 9972–9977. [CrossRef] [PubMed]
3. Morton, S.M.; Silverstein, D.W.; Jensen, L. Theoretical Studies of Plasmonics Using Electronic Structure Methods. *Chem. Rev.* **2011**, *111*, 3962–3994. [CrossRef] [PubMed]
4. Volokitin, Y.; Sinzig, J.; de Jongh, L.J.; Schmid, G.; Vargaftik, M.N.; Moiseevi, I.I. Quantum-Size Effects in the Thermodynamic Properties of Metallic Nanoparticles. *Nature* **1996**, *384*, 621–623. [CrossRef]
5. Dutta, S.; Sankaran, K.; Moors, K.; Pourtois, G.; Van Elshocht, S.; Bömmels, J.; Vandervorst, W.; Tőkei, Z.; Adelmann, C. Thickness Dependence of the Resistivity of Platinum-Group Metal Thin Films. *J. Appl. Phys.* **2017**, *122*, 025107. [CrossRef]
6. Ringe, E.; Sharma, B.; Henry, A.-I.; Marks, L.D.; Duyne, R.P.V. Single Nanoparticle Plasmonics. *Phys. Chem. Chem. Phys.* **2013**, *15*, 4110–4129. [CrossRef] [PubMed]
7. Billaud, P.; Marhaba, S.; Grillet, N.; Cottancin, E.; Bonnet, C.; Lermé, J.; Vialle, J.-L.; Broyer, M.; Pellarin, M. Absolute Optical Extinction Measurements of Single Nano-Objects by Spatial Modulation Spectroscopy Using a White Lamp. *Rev. Sci. Instrum.* **2010**, *81*, 043101. [CrossRef]
8. Andrade, J.D. X-ray Photoelectron Spectroscopy (XPS). In *Surface and Interfacial Aspects of Biomedical Polymers: Volume 1 Surface Chemistry and Physics*; Springer: Boston, MA, USA, 1985; pp. 105–195. ISBN 978-1-4684-8610-0.
9. Cole, K.M.; Kirk, D.W.; Thorpe, S.J. Co_3O_4 Nanoparticles Characterized by XPS and UPS. *Surf. Sci. Spectra* **2021**, *28*, 014001. [CrossRef]
10. Lermé, J.; Palpant, B.; Prével, B.; Pellarin, M.; Treilleux, M.; Vialle, J.L.; Perez, A.; Broyer, M. Quenching of the Size Effects in Free and Matrix-Embedded Silver Clusters. *Phys. Rev. Lett.* **1998**, *80*, 5105–5108. [CrossRef]
11. Khabarov, K.; Nouraldeen, M.; Tikhonov, S.; Lizunova, A.; Efimov, A.; Ivanov, V. Modification of Aerosol Gold Nanoparticles by Nanosecond Pulsed-Periodic Laser Radiation. *Nanomaterials* **2021**, *11*, 2701. [CrossRef]
12. Heiz, U.; Landman, U. *Nanocatalysis*; Springer Science & Business Media: Berlin/Heidelberg, Germany, 2007; ISBN 978-3-540-32646-5.
13. Maier, S.A. *Plasmonics: Fundamentals and Applications*; Springer Science & Business Media: Berlin/Heidelberg, Germany, 2007; ISBN 978-0-387-37825-1.
14. Pichat, P. *Photocatalysis*; MDPI: Basel, Switzerland, 2018; ISBN 978-3-03842-183-2.
15. Ebrahimzadeh, M.A.; Mortazavi-Derazkola, S.; Zazouli, M.A. Eco-Friendly Green Synthesis of Novel Magnetic $Fe_3O_4/SiO_2/ZnO$-Pr_6O_{11} Nanocomposites for Photocatalytic Degradation of Organic Pollutant. *J. Rare Earths* **2020**, *38*, 13–20. [CrossRef]
16. Ebrahimzadeh, M.A.; Mortazavi-Derazkola, S.; Zazouli, M.A. Eco-Friendly Green Synthesis and Characterization of Novel $Fe_3O_4/SiO_2/Cu_2O$–Ag Nanocomposites Using Crataegus Pentagyna Fruit Extract for Photocatalytic Degradation of Organic Contaminants. *J. Mater. Sci. Mater. Electron.* **2019**, *30*, 10994–11004. [CrossRef]
17. Merkl, P.; Long, S.; McInerney, G.M.; Sotiriou, G.A. Antiviral Activity of Silver, Copper Oxide and Zinc Oxide Nanoparticle Coatings against SARS-CoV-2. *Nanomaterials* **2021**, *11*, 1312. [CrossRef]
18. Sirelkhatim, A.; Mahmud, S.; Seeni, A.; Kaus, N.H.M.; Ann, L.C.; Bakhori, S.K.M.; Hasan, H.; Mohamad, D. Review on Zinc Oxide Nanoparticles: Antibacterial Activity and Toxicity Mechanism. *Nano-Micro Lett.* **2015**, *7*, 219–242. [CrossRef]
19. Mohammadzadeh, P.; Shafiee Ardestani, M.; Mortazavi-Derazkola, S.; Bitarafan-Rajabi, A.; Ghoreishi, S.M. PEG-Citrate Dendrimer Second Generation: Is This a Good Carrier for Imaging Agents In Vitro and In Vivo? *IET Nanobiotechnol.* **2019**, *13*, 560–564. [CrossRef]
20. Anker, J.N.; Hall, W.P.; Lyandres, O.; Shah, N.C.; Zhao, J.; Van Duyne, R.P. Biosensing with Plasmonic Nanosensors. In *Nanoscience and Technology*; Macmillan Publishers Ltd.: London, UK; World Scientific Publishing: London, UK, 2009; pp. 308–319. ISBN 978-981-4282-68-0.
21. Gegusin, J. *Physics of Sintering*; Nauka: Moscow, Russia, 1984.
22. Khabarov, K.; Urazov, M.; Lizunova, A.; Kameneva, E.; Efimov, A.; Ivanov, V. Influence of Ag Electrodes Asymmetry Arrangement on Their Erosion Wear and Nanoparticle Synthesis in Spark Discharge. *Appl. Sci.* **2021**, *11*, 4147. [CrossRef]

23. Lakowicz, J.R. *Principles of Fluorescence Spectroscopy*; Springer Science & Business Media: Berlin/Heidelberg, Germany, 2013; ISBN 978-1-4757-3061-6.
24. Reyes Gómez, F.; Rubira, R.J.G.; Camacho, S.A.; Martin, C.S.; Da Silva, R.R.; Constantino, C.J.L.; Alessio, P.; Oliveira, O.N.; Mejía-Salazar, J.R. Surface Plasmon Resonances in Silver Nanostars. *Sensors* **2018**, *18*, 3821. [CrossRef]
25. Khrustov, V.R.; Ivanov, V.V.; Kotov, Y.A.; Kaigorodov, A.S.; Ivanova, O.F. Nanostructured Composite Ceramic Materials in the ZrO_2-Al_2O_3 System. *Glass Phys. Chem.* **2007**, *33*, 379–386. [CrossRef]
26. Gao, Y.; Zhang, X.; Li, Y.; Liu, H.; Wang, Y.; Chang, Q.; Jiao, W.; Song, Y. Saturable Absorption and Reverse Saturable Absorption in Platinum Nanoparticles. *Opt. Commun.* **2005**, *251*, 429–433. [CrossRef]
27. Krayer, L.J.; Kim, J.; Munday, J.N. Near-Perfect Absorption throughout the Visible Using Ultra-Thin Metal Films on Index-near-Zero Substrates [Invited]. *Opt. Mater. Express* **2019**, *9*, 330–338. [CrossRef]
28. Wu, J.-M.; Wu, C.-H. Sintering Behaviour of Highly Agglomerated Ultrafine Zirconia Powders. *J. Mater. Sci.* **1988**, *23*, 3290–3299. [CrossRef]
29. Zook, J.M.; Rastogi, V.; MacCuspie, R.I.; Keene, A.M.; Fagan, J. Measuring Agglomerate Size Distribution and Dependence of Localized Surface Plasmon Resonance Absorbance on Gold Nanoparticle Agglomerate Size Using Analytical Ultracentrifugation. *Am. Chem. Soc. Nano* **2011**, *5*, 8070–8079. [CrossRef]
30. Oliveira, M.J.; Quaresma, P.; Peixoto de Almeida, M.; Araújo, A.; Pereira, E.; Fortunato, E.; Martins, R.; Franco, R.; Águas, H. Office Paper Decorated with Silver Nanostars—An Alternative Cost Effective Platform for Trace Analyte Detection by SERS. *Sci. Rep.* **2017**, *7*, 2480. [CrossRef]
31. Zaheer, Z. Rafiuddin Multi-Branched Flower-like Silver Nanoparticles: Preparation and Characterization. *Colloids Surf. A Physicochem. Eng. Asp.* **2011**, *384*, 427–431. [CrossRef]
32. Yakubovsky, D.I.; Arsenin, A.V.; Stebunov, Y.V.; Fedyanin, D.Y.; Volkov, V.S. Optical Constants and Structural Properties of Thin Gold Films. *Opt. Express* **2017**, *25*, 25574–25587. [CrossRef] [PubMed]
33. Alrahili, M.; Savchuk, V.; McNear, K.; Pinchuk, A. Absorption Cross Section of Gold Nanoparticles Based on NIR Laser Heating and Thermodynamic Calculations. *Sci. Rep.* **2020**, *10*, 18790. [CrossRef] [PubMed]
34. Gong, J.; Dai, R.; Wang, Z.; Zhang, Z. Thickness Dispersion of Surface Plasmon of Ag Nano-Thin Films: Determination by Ellipsometry Iterated with Transmittance Method. *Sci. Rep.* **2015**, *5*, 9279. [CrossRef] [PubMed]

 *materials*

Communication

Surface Texturing of Polyethylene Terephthalate Induced by Excimer Laser in Silver Nanoparticle Colloids

Jakub Siegel [1,*], Tatiana Savenkova [1], Jana Pryjmaková [1], Petr Slepička [1], Miroslav Šlouf [2] and Václav Švorčík [1]

[1] Department of Solid State Engineering, University of Chemistry and Technology Prague, 166 28 Prague, Czech Republic; savenkot@vscht.cz (T.S.); pryjmakj@vscht.cz (J.P.); slepickp@vscht.cz (P.S.); svorcikv@vscht.cz (V.Š.)

[2] Institute of Macromolecular Chemistry, Academy of Sciences of the Czech Republic, Heyrovského nám. 2, 162 06 Prague, Czech Republic; slouf@imc.cas.cz

* Correspondence: jakub.siegel@vscht.cz; Tel.: +420–220–445–149

Abstract: We report on a novel technique of surface texturing of polyethylene terephthalate (PET) foil in the presence of silver nanoparticles (AgNPs). This approach provides a variable surface morphology of PET evenly decorated with AgNPs. Surface texturing occurred in silver nanoparticle colloids of different concentrations under the action of pulse excimer laser. Surface morphology of PET immobilized with AgNPs was observed by AFM and FEGSEM. Atomic concentration of silver was determined by XPS. A presented concentration-controlled procedure of surface texturing of PET in the presence of silver colloids leads to a highly nanoparticle-enriched polymer surface with a variable morphology and uniform nanoparticle distribution.

Keywords: polymer; silver nanoparticles; surface texturing; laser; surface morphology

Citation: Siegel, J.; Savenkova, T.; Pryjmaková, J.; Slepička, P.; Šlouf, M.; Švorčík, V. Surface Texturing of Polyethylene Terephthalate Induced by Excimer Laser in Silver Nanoparticle Colloids. *Materials* **2021**, *14*, 3263. https://doi.org/10.3390/ma14123263

Academic Editor: Alina A. Manshina

Received: 31 May 2021
Accepted: 10 June 2021
Published: 12 June 2021

Publisher's Note: MDPI stays neutral with regard to jurisdictional claims in published maps and institutional affiliations.

1. Introduction

Natural and synthetic polymeric materials are widely used in the manufacture of biomedical supplies [1–3]. Due to the versatility of their surface properties, which may easily be tailored by means of physical and chemical processes, polymers are suitable candidates for biomimetic structures in medical applications. These biomaterials are particularly used in tissue engineering [1], implants [4], drug-delivery carriers [5], wound healing materials or in biological imaging [6].

A common problem in current research of polymers for biomedical applications is the adhesion of bacteria and the formation of biofilms on the surface of medical devices. Bacteria cause infections and inflammation. To improve the bactericidal and anti-biofilm properties of biomaterials, various methods of their surface modification have been developed. The most common one is the coating of their surface with antibiotics. However, bacteria succeed in developing mechanisms to neutralize the effect of these therapeutics [7]. According to the World Health Organization, bacterial resistance to antibiotics, which is constantly increasing in all parts of the world, is today one of the greatest threats to health, agriculture and the food industry [8]. In addition, antibiotics have a number of disadvantages, including poor solubility, instability, and side effects. Therefore, alternatives to antibacterial therapy are intensively sought.

Antibacterial coatings of biocompatible polymers based on metal nanoparticles (NPs), especially gold, silver, and copper, have great potential in preventing or reducing the accumulation of biofilms on the surface of medical devices [9,10]. In particular, especially silver nanoparticles (AgNPs) are important candidates for the role of an alternative to conventional antibiotics. Their antimicrobial effect is associated with the release of silver Ag^+ ions and the ability to interact with cell membranes [11]. Zerovalent AgNPs or inorganic silver compounds ionize in the presence of water, body fluids, or exudates. Formed silver ions are biologically active and can react with proteins, amino acid residues, as well as free anions and receptors located on eukaryotic cell membranes. The action of silver on bacteria and fungi is related to its passage through the cell wall and its ability to interact and irreversibly denature key enzyme systems [12]. Within the microorganism, silver generates free radicals that stop important metabolic processes in the cell, e.g., block electron transfer between respiratory chain enzymes, and enter into reactions with the thiol groups of oxidative enzymes. As a result of the disruption of vital functions, the cell gradually dies out [13].

In this work, we followed and deepened our previously published research in the field of laser-induced anchoring of silver NPs on polymers and demonstrated an effective way of tailoring the surface morphology of a polymeric carrier while its simultaneous enriching with AgNPs [14]. This research confirms that the formerly proposed mechanism of silver NP immobilization in polymers is correct and that a higher concentration of silver NPs in the immobilization solution has the same effect as decreasing the fluence of excitation radiation, due to an absorption effect in silver nanoparticle colloids of an increasing Ag concentration. This effect was demonstrated in a set of immobilization solutions, which varied in Ag concentrations ranging from 5 to 30 mg L^1.

2. Materials and Methods

The PET samples (thickness 50 μm, supplied by Goodfellow Cambridge Ltd., Huntingdon, UK) were immersed in a colloid solution of silver nanoparticles (AgNPs) and irradiated by linearly polarized light from KrF excimer laser (fluence 18 mJ cm^{-2}), according to the procedure published in [14].

The concentration of Ag in AgNPs colloids was determined by atomic absorption spectroscopy (Varian Inc., Palo Alto, CA, USA). Ultraviolet–visible spectroscopy (PerkinElmer Inc., Waltham, MA, USA) was used to study the optical properties of colloidal dispersions of AgNPs. AgNPs were visualized by transmission electron microscopy. Surface morphology and roughness was measured by atomic force microscope. Surface morphology of the samples was also analyzed by a high-resolution FEGSEM microscope (TESCAN, Brno, Czech Republic). Atomic concentration of Ag (Ag3d) on the surface of the samples was determined by X-ray photoelectron spectroscopy (Omicron Nanotechnology GmbH, Taunusstein, Germany). Instrumentation, measurement settings, detailed experimental conditions and measurement errors of individual analyzes are detailed in [14].

3. Results and Discussion

Figure 1 shows a typical TEM image of as-synthesized AgNPs colloids, the inset displays used AgNPs colloids of decreasing concentration (in mg L^{-1}). Indeed, our synthesis procedure provides round-shape silver nanoparticles with a fairly narrow-size distribution. The concentration of Ag in as-synthesized AgNPs colloids was 45 mg L^{-1}. As-synthesized AgNPs were diluted by buffer solution (0.4 mM sodium citrate) for irradiation purposes with 5, 10, 15, 20, 25, and 30 mg L^{-1}. After diluting the solutions to the desired concentrations, UV–Vis analysis was performed (Figure 2) to verify the colloid stability. UV–Vis spectroscopy is often used to estimate the size, shape, and particle distribution of colloid solutions of metal nanoparticles since they show a specific absorption band corresponding to localized surface plasmon resonance (LSPR) [15]. Regarding the composition of AgNPs solution (0.4 mM of sodium citrate in water), the specific shape of the absorption bands truly corresponds to round-shaped NPs of narrow-size distribution and a mean diameter around 25 nm [16], which corresponds well with the TEM analysis. The position of the LSPR maxima at 409 nm remains practically unchanged in the case of all examined solutions which points to particle aggregation resistance and a good colloid stability. A recorded increase in intensity corresponds to the increasing concentration in AgNPs.

Immediately after PET irradiation in AgNPs colloids, the samples were rinsed by distilled water, dried with a nitrogen stream, and placed into a desiccator for 24 h. Surface morphology of pristine and AgNPs immobilized PET was studied by AFM and is depicted in Figure 3.

Figure 1. TEM image of as-synthesized silver nanoparticles. Inset shows AgNPs colloids with a decreasing Ag concentration (numbers stand for Ag concentration in mg L^{-1}).

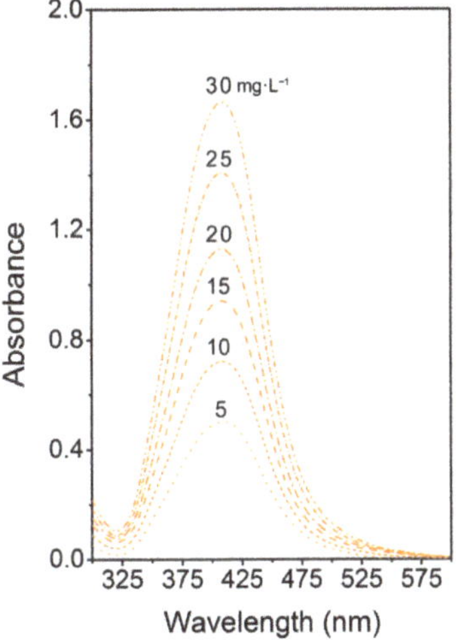

Figure 2. UV–Vis absorption spectra of AgNPs colloids of different Ag concentrations.

Pristine PET exhibits a flat surface with surface roughness Ra about 1 nm (see Figure 4, blue bar), which is typical for commercially available foils. Once PET is laser irradiated in the presence of AgNPs colloids, the surface morphology changes dramatically, corresponding to AgNPs incorporation to the polymer surface and transformation of the surface morphology due to light absorption. The development of the surface morphology of the polymer is in accordance with the immobilization mechanism of NPs by the action of intense light of wavelengths near the maximum of their plasmon resonance, described in detail in our previous work [14]. As the Ag concentration in the immobilization solution increases, a proportional dose of light energy is absorbed due to plasmon excitation in AgNPs, which reduces the part of the energy absorbed by the polymer surface and, thus, prevents a change in its surface morphology. Thus, PET immobilized at AgNPs colloids of higher concentration (30 mg L^{-1}) contains the highest amount of silver (see Figure 4, orange bar) while its surface morphology is quite flat, reminding that of a pristine polymer. On the other hand, when the immobilization solution contains the lowest concentration of

AgNPs (5 mg L^{-1}), surface roughness increases significantly (see Figure 4, blue bar) and the detected amount of Ag drops down (Figure 4, orange bar).

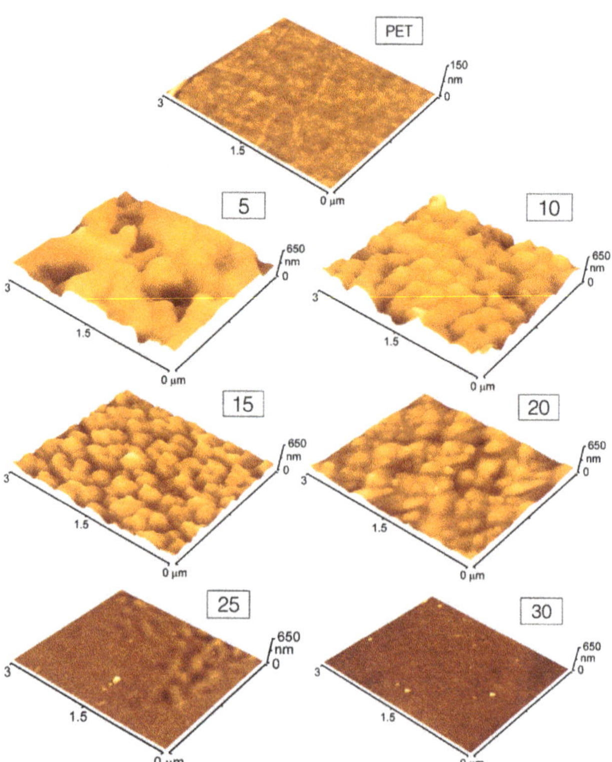

Figure 3. AFM images of PET after its irradiation in the presence of AgNPs colloids of different Ag concentrations. Numbers in the boxes indicate the concentration of silver in the immobilization solution. (Notice: Z-axis scaling in the case of pristine PET is limited to 150 nm due to naturally flat surface morphology).

We completed an AFM observation with FEGSEM analysis (Figure 5) to better visualize AgNPs on the surface of PET. Figure 5 shows corrugated and smooth PET foils with Ag nanoparticles immobilized in colloids with a Ag concentration of 15 and 30 mg L^{-1}, respectively. Presented micrographs illustrate that the morphology is quite uniform, with evenly distributed AgNPs over the polymer surface. It is evident that the irradiation in higher concentrated AgNPs solution resulted in a smooth surface with planar morphology decorated with nanoparticles, while under lower concentration a rough and rugged polymer surface was obtained. Those findings are in good accordance with AFM. They also point to the key issue of concentration-controlled immobilization compared to the laser fluence-controlled approach, presented in our former study [14]. The concentration-controlled approach benefits from the application of constant laser fluence, which strictly preserves its effect on particle size through the immobilization process. Thus, the immobilized particles under the concentration-controlled approach are of the same size regardless of the resulting polymer morphology (smooth or rough). This does not apply in the case of the fluence-controlled approaches, where an increase in laser fluence (leading to a rough surface morphology) causes an increase in particle size due to considerably higher absorbed energy and promoted particle ingrowth [17].

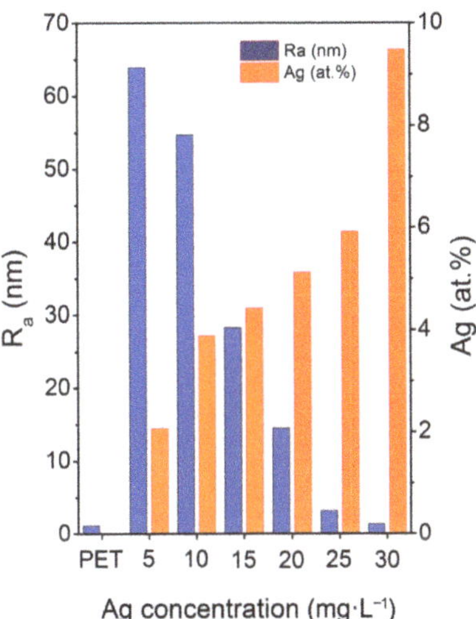

Figure 4. Average surface roughness (Ra) and atomic concentration of silver determined by XPS as a function of the Ag concentration in the immobilization solution.

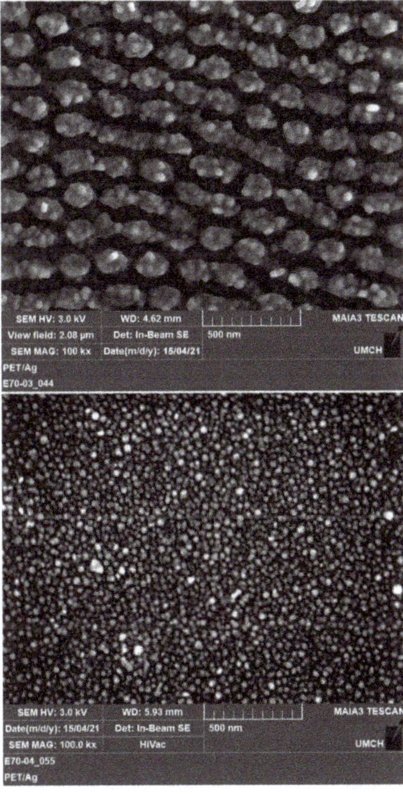

Figure 5. FEGSEM micrographs showing the surface morphology of PET after its irradiation in the presence of AgNPs colloids of different Ag concentrations: 15 mg L^{-1} (up) and 30 mg L^{-1} (down).

It is also worth emphasizing that, as the Ag concentration in the immobilization solution increases, the silver concentration over the PET surface increases proportionally (see Figure 4), while the R_a value monotonously decreases as the surface morphology becomes smoother owing to the fact that more incoming energy is absorbed by the particles.

4. Conclusions

In this work we introduced a simple and versatile approach for laser induced texturing of PET foil in the presence of silver nanoparticles. This method is based on laser processing of polymers in the presence of AgNPs colloids of variable concentration. Compared to laser fluence-controlled approaches, our procedure enables to maintain a size of nanoparticles while controlling the resulting morphology of the polymer. We demonstrated that within the AgNPs concentration range of 5–30 mg L^{-1}, where a rough and smooth polymer morphology may be produced, the PET foil was evenly decorated by silver nanoparticles. An increasing concentration of silver in the immobilization solution leads to a monotonic increase in the concentration of silver detected on the polymer surface. Surface texturing of polymers highly enriched with silver nanoparticles can create new opportunities in the development of prospective antimicrobial coatings, especially suitable for inhibition of biofilm formation.

Author Contributions: Conceptualization, methodology, funding acquisition, writing the manuscript, J.S.; sample preparation, nanoparticle immobilization, T.S. and J.P.; AFM analysis, P.S.; SEM analysis and data evaluation, M.Š.; data curation, V.Š. All authors have read and agreed to the published version of the manuscript.

Funding: This research was funded by the Czech Science foundation, grant number 21-05506S and OP VVV Project NANOTECH ITI II, grant number CZ.02.1.01/0.0/0.0/18_069/0010045.

Institutional Review Board Statement: Not applicable.

Informed Consent Statement: Not applicable.

Data Availability Statement: The data presented in this study are available on request from the corresponding author.

Acknowledgments: J.S. is thankful to Adéla Siegelová for personal support during the experimental work and manuscript preparation.

Conflicts of Interest: The authors declare no conflict of interest.

References

1. Pryjmakova, J.; Kaimlova, M.; Hubacek, T.; Svorcik, V.; Siegel, J. Nanostructured Materials for Artificial Tissue Replacements. *Int. J. Mol. Sci.* **2020**, *21*, 2521. [CrossRef] [PubMed]
2. Slepicka, P.; Siegel, J.; Lyutakov, O.; Kasalkova, N.S.; Kolska, Z.; Bacakova, L.; Svorcik, V. Polymer nanostructures for bioapplications induced by laser treatment. *Biotechnol. Adv.* **2018**, *36*, 839–855. [CrossRef] [PubMed]
3. Polivkova, M.; Strublova, V.; Hubacek, T.; Rimpelova, S.; Svorcik, V.; Siegel, J. Surface characterization and antibacterial response of silver nanowire arrays supported on laser-treated polyethylene naphthalate. *Mater. Sci. Eng. C Mater. Biol. Appl.* **2017**, *72*, 512–518. [CrossRef] [PubMed]
4. Teo, A.J.T.; Mishra, A.; Park, I.; Kim, Y.-J.; Park, W.-T.; Yoon, Y.-J. Polymeric Biomaterials for Medical Implants and Devices. *ACS Biomater. Sci. Eng.* **2016**, *2*, 454–472. [CrossRef] [PubMed]
5. Mora-Huertas, C.E.; Fessi, H.; Elaissari, A. Polymer-based nanocapsules for drug delivery. *Int. J. Pharm.* **2010**, *385*, 113–142. [CrossRef] [PubMed]
6. Yu, J.; Rong, Y.; Kuo, C.-T.; Zhou, X.-H.; Chiu, D.T. Recent Advances in the Development of Highly Luminescent Semiconducting Polymer Dots and Nanoparticles for Biological Imaging and Medicine. *Anal. Chem.* **2017**, *89*, 42–56. [CrossRef] [PubMed]
7. Skarzynska, M.; Zajac, M.; Wasyl, D. Antibiotics and bacteria: Mechanisms of action and resistance strategieS. *Adv. Microbiol.* **2020**, *59*, 49–62.
8. Mahase, E. More countries report on antibiotic resistance but results are "worrying," says WHO. *BMJ Brit. Med. J.* **2020**, *369*, m2217. [CrossRef] [PubMed]
9. Eleraky, N.E.; Allam, A.; Hassan, S.B.; Omar, M.M. Nanomedicine Fight against Antibacterial Resistance: An Overview of the Recent Pharmaceutical Innovations. *Pharmaceutics* **2020**, *12*, 142. [CrossRef] [PubMed]

10. Abdalla, S.S.I.; Katas, H.; Azmi, F.; Busra, M.F.M. Antibacterial and Anti-Biofilm Biosynthesised Silver and Gold Nanoparticles for Medical Applications: Mechanism of Action, Toxicity and Current Status. *Curr. Drug Del.* **2020**, *17*, 88–100. [CrossRef] [PubMed]
11. Polivkova, M.; Hubacek, T.; Staszek, M.; Svorcik, V.; Siegel, J. Antimicrobial Treatment of Polymeric Medical Devices by Silver Nanomaterials and Related Technology. *Int. J. Mol. Sci.* **2017**, *18*, 419. [CrossRef] [PubMed]
12. Lansdown, A.B.G. Silver in health care: Antimicrobial effects and safety in use. *Curr. Probl. Dermatol.* **2006**, *33*, 17–34. [PubMed]
13. Yang, G.; Xie, J.; Hong, F.; Cao, Z.; Yang, X. Antimicrobial activity of silver nanoparticle impregnated bacterial cellulose membrane: Effect of fermentation carbon sources of bacterial cellulose. *Carbohydr. Polym.* **2012**, *87*, 839–845. [CrossRef]
14. Siegel, J.; Kaimlová, M.; Vyhnálková, B.; Trelin, A.; Lyutakov, O.; Slepička, P.; Švorčík, V.; Veselý, M.; Vokatá, B.; Malinský, P.; et al. Optomechanical Processing of Silver Colloids: New Generation of Nanoparticle–Polymer Composites with Bactericidal Effect. *Int. J. Mol. Sci.* **2021**, *22*, 312. [CrossRef] [PubMed]
15. Amendola, V.; Bakr, O.M.; Stellacci, F. A Study of the Surface Plasmon Resonance of Silver Nanoparticles by the Discrete Dipole Approximation Method: Effect of Shape, Size, Structure, and Assembly. *Plasmonics* **2010**, *5*, 85–97. [CrossRef]
16. Agnihotri, S.; Mukherji, S.; Mukherji, S. Size-controlled silver nanoparticles synthesized over the range 5–100 nm using the same protocol and their antibacterial efficacy. *RSC Adv.* **2014**, *4*, 3974–3983. [CrossRef]
17. Popov, A.K.; Brummer, J.; Tanke, R.S.; Taft, G.; Loth, M.; Langlois, R.; Wruck, A.; Schmitz, R. Synthesis of isolated silver nanoparticles and their aggregates manipulated by light. *Laser Phys. Lett.* **2006**, *3*, 546–552. [CrossRef]

Article

Effect of Laser Shock Peening on Fretting Fatigue Life of TC11 Titanium Alloy

Xufeng Yang, Hongjian Zhang *, Haitao Cui and Changlong Wen

College of Energy and Power Engineering, Nanjing University of Aeronautics and Astronautics,
Nanjing 210016, China; yangxufeng@nuaa.edu.cn (X.Y.); cuiht@nuaa.edu.cn (H.C.);
changlongwen91@163.com (C.W.)
* Correspondence: zhanghongjian@nuaa.edu.cn; Tel.: +86-181-6810-1903

Received: 2 September 2020; Accepted: 21 October 2020; Published: 22 October 2020

Abstract: The purpose of this paper is to investigate the performance of laser shock peening (LSP) subjected to fretting fatigue with TC11 titanium alloy specimens and pads. Three laser power densities (3.2 GW/cm^2, 4.8 GW/cm^2 and 6.4 GW/cm^2) of LSP were chosen and tested using manufactured fretting fatigue apparatus. The experimental results show that the LSP surface treatment significantly improves the fretting fatigue lives of the fretting specimens, and the fretting fatigue life increases most when the laser power density is 4.8 GW/cm^2. It is also found that with the increase of the laser power density, the fatigue crack initiation location tends to move from the surface to the interior of the specimen.

Keywords: fretting fatigue; laser shock peening; laser power density; TC11 alloy

---

1. Introduction

Fretting is usually recognized as a kind of near-surface damage arising from the relative slip between the two mating components that are clamped together with a normal force [1]. If the relative slip of contact surfaces is caused by the cyclical stress load of one of the contact components, then the crack may nucleate and continue to propagate to fatigue failure, which is called fretting fatigue. Fretting is a problem in many aerospace applications such as airframe lap joints and the dovetail attachment between the engine blade and disk [2]. In fact, fretting or fretting fatigue has been identified as one of the costliest sources of in-service damage in the US Air Force [3]. Considering that the original design configuration of many components is difficult to change, the research on surface protection techniques of fretting fatigue has been developed since the 1960s, and some techniques have been applied successfully [4]. The principle of most surface protection techniques, such as coating and shot peening, is to introduce residual compressive stress into the contact surface, reduce the friction coefficient of the contact surface, and improve the hardness and wear resistance of the contact surface [5–10]. As an advanced surface treatment, laser shock peening (LSP) which induces the deep compressive residual stresses beneath the contact surfaces has been proved to be significantly beneficial to improve the fretting fatigue lives [11].

The simple process of LSP is shown schematically in Figure 1. The region to be treated is covered by an absorbing layer, over which is the confinement layer which is usually used as running water [12]. A high-energy laser pulse is directed on to the surface, which passes through the confinement layer and is absorbed by the absorbing layer which is usually used as a black coating, metal sacrifice layer or black tape [13–15]. Then, the plasma is formed over the absorbing surface. The expansion of the plasma is constrained by the running water, which drives a shockwave into the material, and this shockwave causes the plastic deformation and residual stresses near surface region. The relevant research results show that under certain conditions, there is a critical value for the thickness of the absorbing layer. When the thickness exceeds the critical value, the non-vaporized part will cause the loss of shock wave.

When the thickness is lower than the critical value, the unabsorbed laser energy will vaporize the metal surface, thus forming rough impact pits and reducing the impact effect [16].

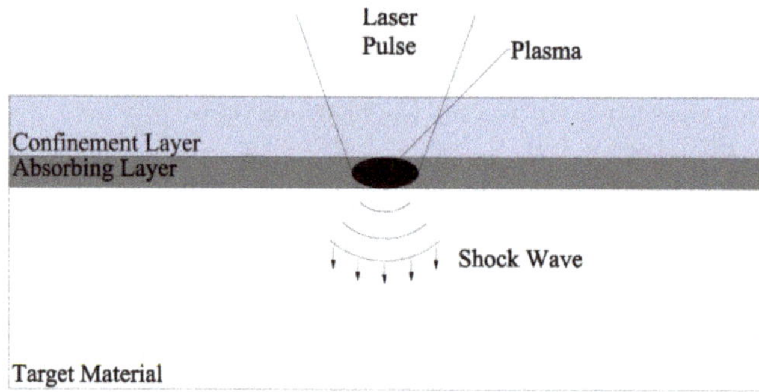

Figure 1. Schematic of the laser shock peening (LSP).

It is believed that compared with the conventional shot peening, LSP can produce compressive stresses to greater depths and has better resistance to fretting fatigue. In fact, the fretting fatigue process can also induce residual compressive stress in the surface layer of materials, and with the increase of the fatigue loading cycle, the residual stress levels presents a trend of first increase and then decrease [17]. However, it is found that fretting fatigue loading on the experimental samples treated with LSP can cause significant stress relaxation extending to a certain depth [18], and this interesting phenomenon may indicate the possibility that the fretting fatigue process may reduce the effect of the LSP. In addition, another interesting result has been found in work on the effect of laser peening and shot peening on fretting fatigue of TC4 titanium alloy [19]. In some fretting fatigue load conditions, the crack initiation points appear in the subsurface layer where the compensated tensile stress is max instead of the surface. It is believed that the reason for the fatigue failure of the specimens is the compensated tensile stress rather than the fretting effect. This result shows that LSP can significantly reduce the impact of fretting fatigue damage, but at the same time, the compensating tensile stress caused by the introduction of residual compressive stress layer should be considered. The superposition of compensating tensile stress and external load will increase the local stress level, which will lead to failure.

Although most of the relevant researches successfully attribute the improvement of fretting fatigue life to the introduction of residual compressive stress influencing layer, few researches care about the influence of the parameters of the laser shock itself. In fact, there are many parameters affecting the fretting fatigue resistance of LSP, including the overlap ratio of laser spots, laser energy, laser spot diameter, impact times, laser pulse duration and so on [20]. Thus, different combination of these parameters may cause different residual stress influencing layers and different effect on fretting fatigue resistance.

Laser power density I_0 can be described as Equation (1) [21], where E is the pulse energy, R means the radius of the laser spot and τ is the laser pulse duration. It is recognized that a better residual compressive stress distribution can be obtained by selecting the appropriate laser power density [21,22], and an excessive laser power density will lead to indentation on the surface of the material, and even ablation of the absorbing layer [23,24].

$$I_0 = \frac{E}{4\pi R^2 \tau} \tag{1}$$

Laser pulse duration τ is used to measure the time distribution characteristics of pulse laser, which has an important influence on the depth of the residual stress influence layer [25]. It is found that with the increase of pulse duration, the duration of shock wave pressure increases, and the plastic deformation of the material intensifies, which is beneficial to improve the surface residual stress after impact strengthening [26–29]. However, when the pulse duration width exceeds a certain value (about

10 ns) [28], the improvement effect of residual stress field in depth direction is relatively stable, and when the pulse duration width continues to increase to 24 ns [27], the surface residual compressive stress decreases. In addition, the excessive pulse duration may lead to the excessive loss of absorbing protective layer and laser ablation of surface materials [27].

In this paper, it was desired to investigate the influence of the laser power density for the effect of laser shock peening on fretting fatigue of TC11 titanium alloy, and three laser power densities of LSP were chosen and tested using manufactured fretting fatigue apparatus.

2. Experiments

2.1. Specimen Materials

The material used here, both for specimens and fretting pads, was TC11 (Ti-6.5Al-1.5Zr-3.5Mo-0.3Si) titanium alloy. TC11 titanium alloy is widely used in the China aviation field, and the chemical composition of this alloy (in wt.%) is given in Table 1 [30]. TC11 titanium alloy is a kind of $\alpha + \beta$ type heat-resisting titanium alloy with an outstanding combination property. It is mainly used to manufacture compressor blades and disks in aero-engines. The heat treatment was double annealing, 950 °C/2 h air cooled and 530 °C/6 h air cooled. The material properties of this alloy at room temperature are listed in Table 2 [30].

Table 1. Chemical composition of titanium alloy.

Composition	Al	Mo	Zr	Si	Fe	Ti
Percent (wt./%)	6.40	3.57	1.63	0.25	0.13	Bal

Table 2. Material parameters for TC11 at room temperature.

E (MPa)	ν	σ_{-1} (MPa)	σ_b (MPa)	$\sigma_{0.2}$ (MPa)
120,000	0.33	540	1139	866

2.2. Laser Shock Peening (LSP) Specimen and Procedure

All the main fretting specimens and the fretting pads were manufactured from the TC11 plate. The fretting pads were flat pads with rounded corners and loaded against the main "dog bone" specimen. The details of the main specimen and pad for fretting fatigue tests are shown in Figure 2. The LSP region, which means laser shock peening area, is a rectangular region of a size of 25 mm × 25 mm, and both two sides of the specimen are under LSP.

The LSP experiment was carried out by the YD60-M165 laser system with a wavelength of 1064 nm and laser pulse width of 20 ns (Figure 3). The surfaces of the test samples were covered with the black tape (0.1 mm thick), which was used as the absorbing layer for plasma initiation to protect the test surface from the thermal damage of high-temperature plasma. During the LSP, the test specimens were under a water bath, and the water layer with a thickness of about 1 mm was used as the confinement layer. During the experiment, the test piece moved in a presupposition route by the manipulator, and the laser beam was fixed and perpendicular to the sample surface all the time.

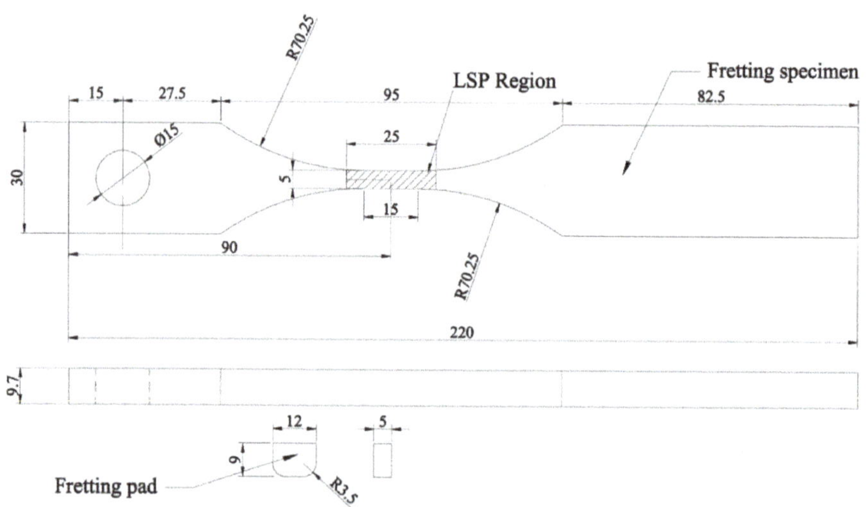

Figure 2. Main specimen and pad for fretting fatigue tests (mm).

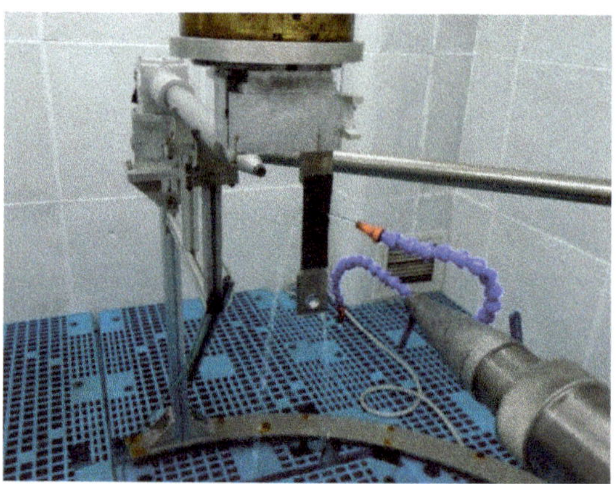

Figure 3. Picture of YD60-M165 laser system.

To guarantee the quality of LSP, multi LSP impacts were used, and the black tape was replaced after each impact to prevent the thermal damage of the sample surface because of the erosion of the black tape. The diameter of the laser spot was 2 mm and the detailed laser sweep steps and direction are shown in Figure 4, and the detailed surface condition after the LSP impacts is shown in Figure 5. The overlapping rate was 50% between two adjacent spots to ensure a good surface condition.

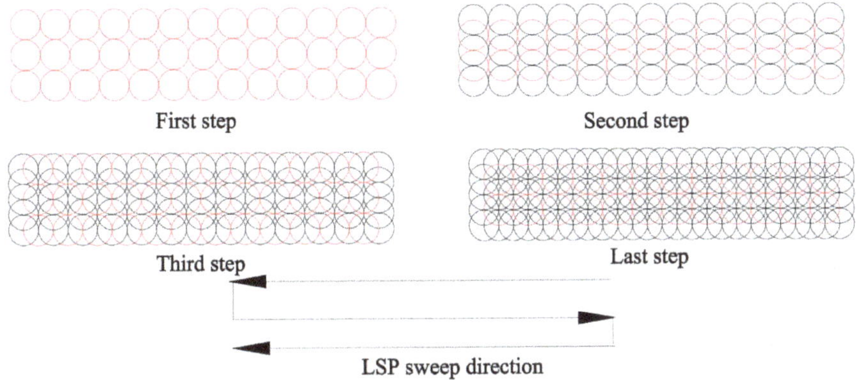

Figure 4. Diagram of the laser sweep steps and direction.

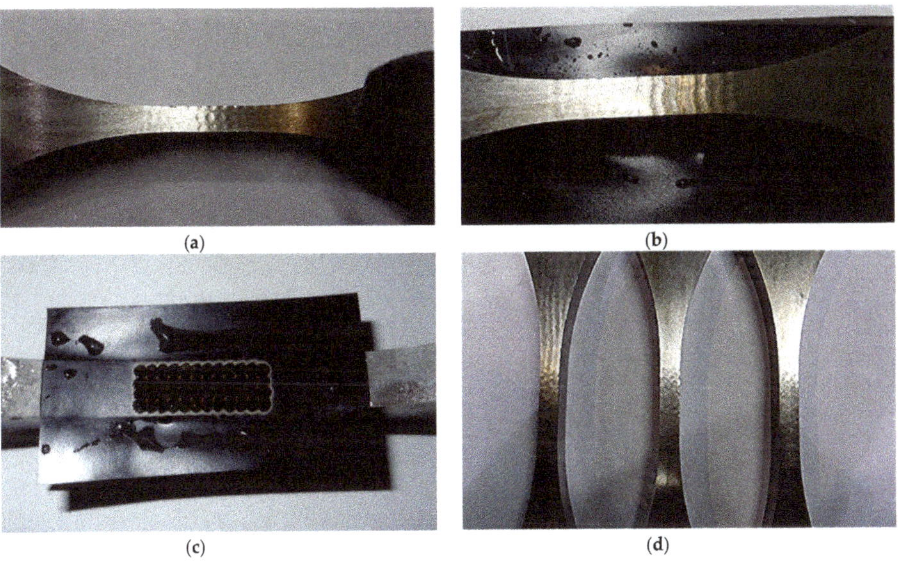

Figure 5. Surface condition after the LSP impacts. (**a**) First step. (**b**) Second step. (**c**) Third step. (**d**) Last step.

The laser power densities chosen here were: 3.2 GW/cm^2, 4.8 GW/cm^2 and 6.4 GW/cm^2. The details of the parameters of laser shock peening are listed in Table 3.

Table 3. Laser shock peening parameters for TC11 specimens.

Wave Length	Pulse Width	Diameter of Laser Spot	Overlapping Rate	Absorbing Layer	Restraint Layer	Laser Power Densities (GW/cm^2)
1064 nm	20 ns	2 mm	50%	Black tape	Water	3.2, 4.8, 6.4

2.3. Surface Topography and Residual Stress

The surface topography of the untreated and LSP samples were carried out by using a non-contact 3D optical profilometer. The Z-direction scanning range of the profilometer is 0.1 nm to 10 mm. The measuring area of the surface topography was 4.5 mm × 3.4 mm, which was larger than the laser spot diameter used in this experiment. This measuring area could contain the typical characteristics of impact surface, and meet the requirements of surface topography and roughness measurement in this paper.

In this paper, the X-ray diffraction (XRD) method is used to test the residual stress, and the test equipment used here is a proto LXRD X-ray diffractometer. The basic principle of X-ray residual stress measurement is shown in Figure 6.

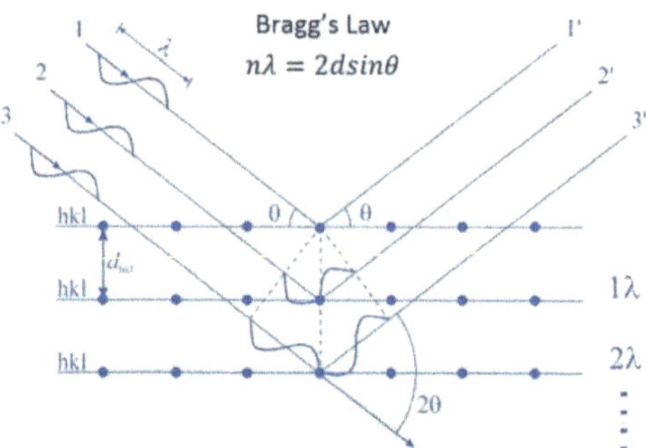

Figure 6. Principle of X-ray residual stress measurement.

When the surface of the metal, which has residual stress, is measured, the crystal lattice spacing d changes. When the X-ray with wavelength λ is incident at angle θ, the corresponding diffraction angle 2θ is measured. The relationship between the optical path difference of X-ray and wavelength λ is as follows:

$$2d \sin \theta = n\lambda \tag{2}$$

The radiation length used in this paper was 1.5418 Å, the diffraction type was Cu-K-Alpha, the diffraction plane (h, k, l) was (2, 1, 3), the crystal structure was HCP and the diffraction angle were 142 degrees. The principle of residual stress measurement used here was the $\sin^2\psi$ method, and the range of the ψ were −39 to +39 degrees. The spring constants used here for the stress calculations were: $S_1 = -2.97 \times 10^{-6}$ MPa^{-1}, $S_2 = 23.78 \times 10^{-6}$ MPa^{-1}. The typical example of the d (interplanar spacing) − $\sin^2\psi$ diagram was shown in Figure 7.

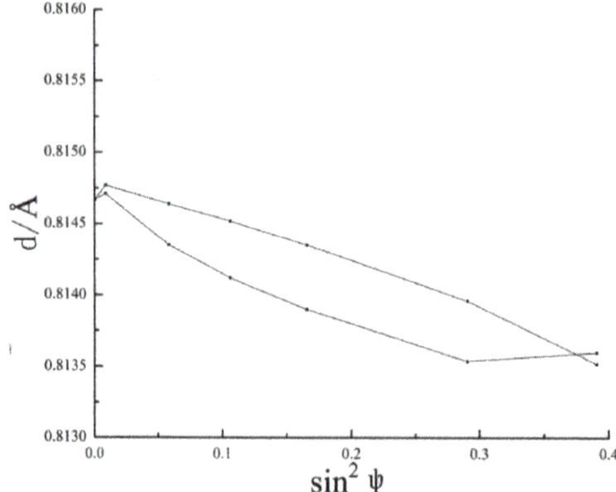

Figure 7. Diagram of the d–$\sin^2\psi$.

For the convenience of the measurement, the test sample used here is shown in Figure 8. All the five test points are on the midline of the test sample, and the test point 3 is at the center of the LSP region. The LSP region, which contains test points 2, 3, and 4, is in the center area of the test sample. Test point 1 and 5 are out of the LSP region and reflect the residual stress of the untreated area.

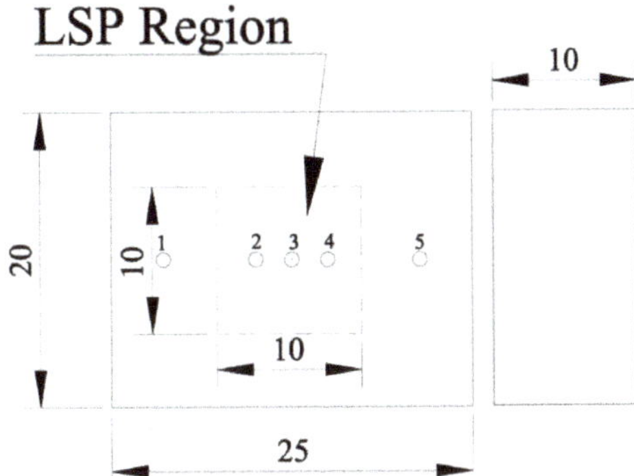

Figure 8. Test sample for surface residual stress (mm).

Due to the limited penetration of X-ray, the residual stress on the surface of the material could only be measured. In view of the residual stress inside the material, the electrolytic polishing method was used to peel off the surface material layer by layer, and the polishing fluid was 24% HNO_3, 14% HF and 62% H_2O (volume ratio). The diameter of the etched area was 10 mm, and each period of electro polishing was 15 s, and the total polishing measurement was taken five times. The thickness of the material layer removed by etching was measured after each time of electro polishing. Then, the residual stress inside of the material could be measured by proto LXRD X-ray diffractometer.

2.4. Fretting Fatigue Test

To conduct the fretting tests, the fretting fatigue test rig, which is constructed by an ordinary load frame with an additional pad fixture and a hydraulic cylinder, was used. A schematic diagram of the fretting fatigue rig is shown in Figure 9.

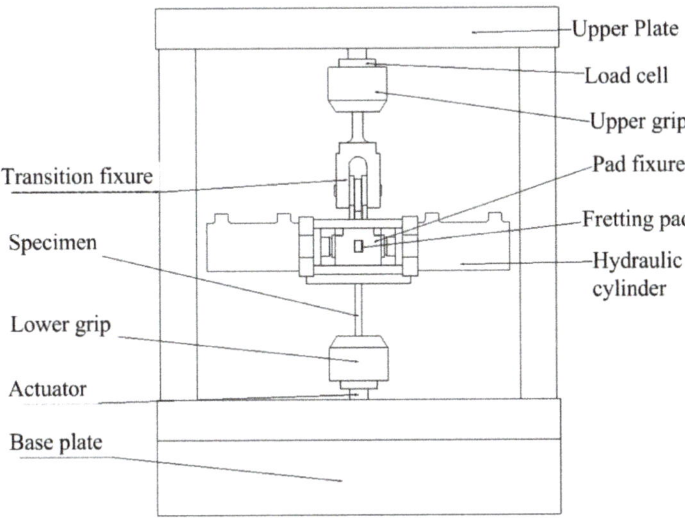

Figure 9. Diagram of the fretting fatigue test rig.

The bottom of the main specimen is held by the moveable lower grip, while the top of the specimen is connected with the pad fixture and the transition fixture, which is held by the fixed upper grip, through a bolted connection. The moveable lower grip is mounted to the load cell and to a hydraulic actuator capable of applying fatigue loads to the specimen up to 50 KN. The two fretting pads are then held by the pad fixture, which is shown in Figure 10, so that the pads can be clamped by the hydraulic cylinder and contacted with the specimen. The hydraulic cylinder, which produces the normal load to the specimen, is fixed on the pad fixture with four long bolts, and to keep this normal load constant through the whole test, an accumulator is connected to the hydraulic line. The normal load is then calculated from the hydraulic pressure.

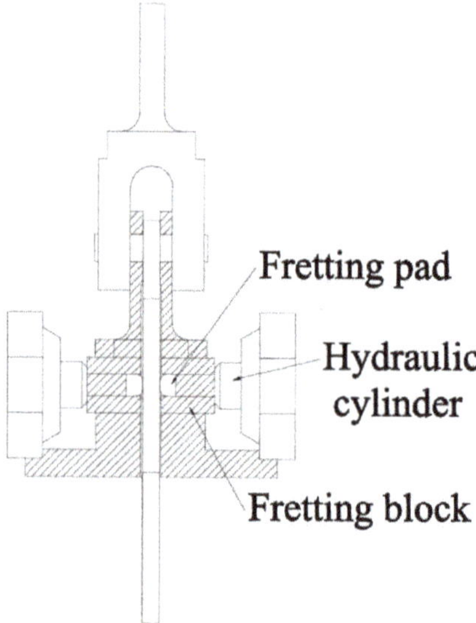

Figure 10. Diagram of the pad fixture.

The static normal force, P, was first applied to the fretting pads. Then the cycle tensile load varying between Q_{min} and Q_{max} was applied to the specimen. Figure 11 shows a simple load configuration of the test. Table 4 reports the experimental parameters of the tests. Eight fretting fatigue tests were carried out and the lives of the crack initiation were recorded. Strain gauges bonded on both sides of the main specimen (Figure 12) were used to monitor the initiation of the crack. Once the crack occurred, the strain curves would rise immediately until the strain gauges were broken. At the same time, the hole in the middle of the pad fixture could help us to observe the crack initiation.

Table 4. Experimental parameters used in fretting fatigue tests.

Series	P (MPa)	Q_{max} (MPa)	Stress Ratio	Laser Power Densities (GW/cm^2)
1	65.45	400	0.1	No treatment
2	65.45	400	0.1	3.2
3	65.45	400	0.1	4.8
4	65.45	400	0.1	6.4
5	40	400	0.1	No treatment
6	40	400	0.1	3.2
7	40	400	0.1	4.8
8	40	400	0.1	6.4

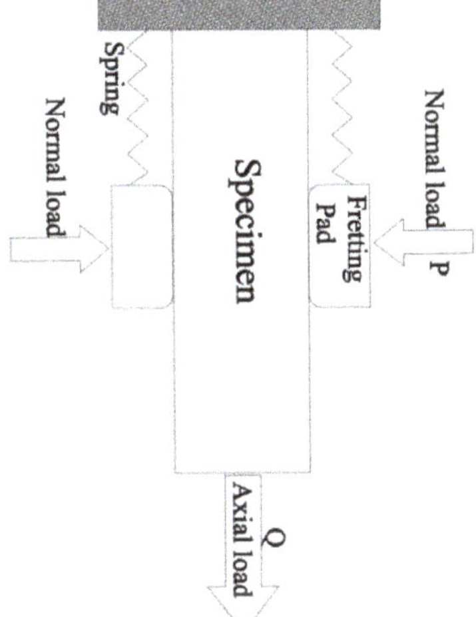

Figure 11. Simple load configuration of the test.

Figure 12. Picture of the strain gauges.

3. Results and Discussion

Three-dimensional surface morphologies and surface roughness of the main specimens with different laser power densities are shown in Figure 13 and Table 5 respectively. There are obvious pits on the surface of the LSP area, and there are convex parts in the adjacent areas of the two pits, and the surface of the specimen is slightly uneven. In this experiment, the overlap rate of adjacent spots was 50% so that all the surfaces of the test area were affected by LSP, and some of them were overlapped areas, where plastic deformation was relatively large. Compared with the untreated samples, the surfaces of the LSP samples have obvious pits and lager surface roughness. The surface roughness increases with the increase of laser power density, which means a larger laser power density may cause a worse surface condition.

Table 6 shows the results of the surface compressive residual stresses of the test samples in the laser power densities of 3.2 GW/cm^2, 4.8 GW/cm^2 and 6.4 GW/cm^2, and the average surface compressive residual stresses of the treated part are −554 MPa, −650 MPa and −645 MPa respectively. It seems that with the increase of the laser power densities, the average surface compressive residual stresses will reach a peak value and change a little. Continuing to increase the laser power density will cause the peak pressure of shock wave to be too high, which will affect the surface quality of the material. In addition, the residual compressive stress on the surface of the material will be reduced to a certain extent due to the generation of a surface unloading wave.

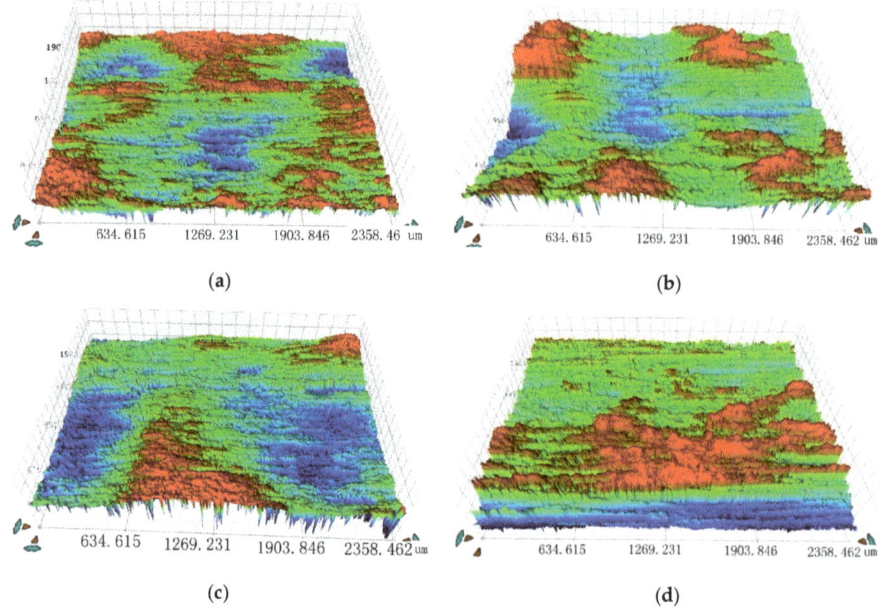

Figure 13. Three-dimensional surface morphologies of the main specimens. (**a**) 3.2 GW/cm^2 (**b**) 4.8 GW/cm^2 (**c**) 6.4 GW/cm^2 (**d**) untreated.

Table 5. Surface roughness of the main specimens with different laser power densities.

Laser Power Density/(GW/cm^2)	0	3.2	4.8	6.4
Roughness/μm	0.94	1.10	1.28	1.52

Table 6. Results of the surface compressive residual stresses.

I_0/(GW/cm^2)	Residual Stress/(MPa)	Point 3	Point 2	Point 1	Point 4	Point 5
3.2	σ_x	−138	−521	−563	−559	−121
	σ_y	−121	−488	−601	−589	−135
4.8	σ_x	−171	−676	−671	−670	−168
	σ_y	−161	−639	−630	−619	−164
6.4	σ_x	−132	−674	−735	−633	−130
	σ_y	−90	−569	−696	−561	−99

Figure 14 shows the distribution of the compressive residual stresses inside of the samples in the laser power densities of 6.4 GW/cm^2, and the residual compressive stress affects the depth of the layer by at least 0.8 mm.

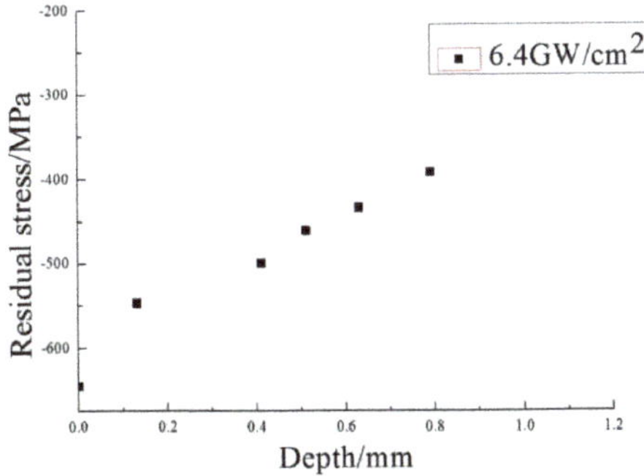

Figure 14. Distribution of the compressive residual stresses inside of the samples (6.4 GW/cm^2).

Table 7 and Figure 15 shows the fretting fatigue life and average fretting fatigue life of all the specimens respectively.

Table 7. Experimental fretting fatigue initiation life of TC11 with different laser power densities.

Load	Series 1 (No Treatment)		Series 2 (I_0 = 3.2 GW/cm^2)	
	Life (Cycles)	Average Life (Cycles)	Life (Cycles)	Average Life (Cycles)
P = 65.45 MPa, Q_{max} = 400 MPa	105,223 82,750 78,616	88,863	152,527 237,018 206,710	198,752
Load	Series 3 (I_0 = 4.8 GW/cm^2)		Series 4 (I_0 = 6.4 GW/cm^2)	
P = 65.45 MPa, Q_{max} = 400 MPa	357,900 333,157 279,899	323,652	195,221 18,823	192,022
Load	Series 5 (No treatment)		Series 6 (I_0 = 3.2 GW/cm^2)	
P = 40 MPa, Q_{max} = 400 MPa	133,233 79,450 104,500 109,214	106,599	>400,000	-
Load	Series 7 (I_0 = 4.8 GW/cm^2)		Series 8 (I_0 = 6.4 GW/cm^2)	
P = 40 MPa, Q_{max} = 400 MPa	>430,000	-	282,828	282,828

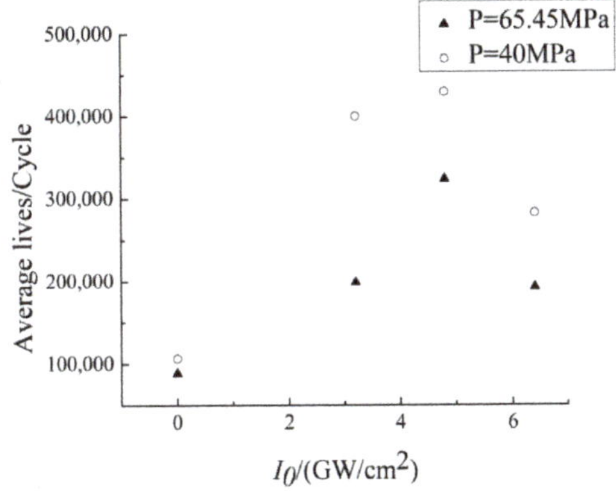

Figure 15. Average fretting fatigue life of TC11 with different laser power densities.

Compared with the untreated specimens, the fretting fatigue lives of LSP specimens have been significant increased, and the laser power density of 4.8 GW/cm^2 shows the best effect to prolong the fretting fatigue life. As mentioned before, the compressive residual stress of the laser power density of 4.8 GW/cm^2 and 6.4 GW/cm^2 is similar, but the surface condition of the laser power density of 6.4 GW/cm^2 is worse, which causes a shorter fatigue life. All the fatigue crack initiation location of the specimen was on the edge of the contact area near the lower grip, and the crack propagation direction was almost perpendicular to the moving direction, which is shown in Figure 16.

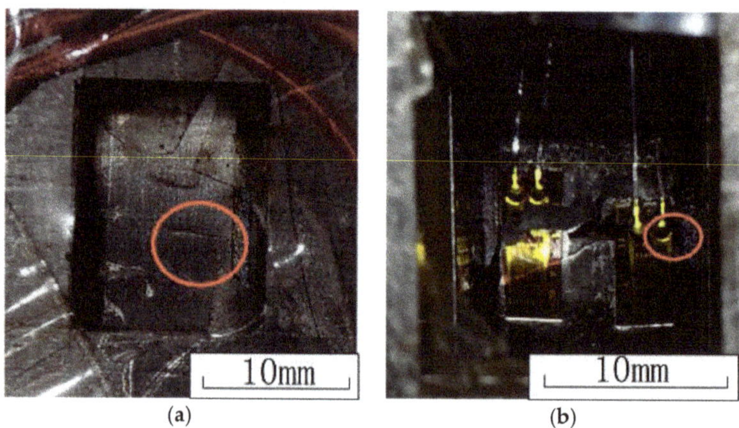

Figure 16. Pictures of the crack initiation and Specimen fracture. (a) Crack initiation. (b) Specimen fracture.

The typical wear surfaces of the fretting pads were taken with an optical microscope (OM) under the same load conditions and different laser power densities are shown in Figure 17. It is clear to see that because of the fretting phenomenon between the specimen and pads, the debris is accumulated on the edge of the contact area in Figure 17a. It is also seen that because of the larger surface roughness caused by LSP impact, there are no remarkable wear traces in some parts of the LSP samples' sliding area, which is similar to Kevin's research [19].

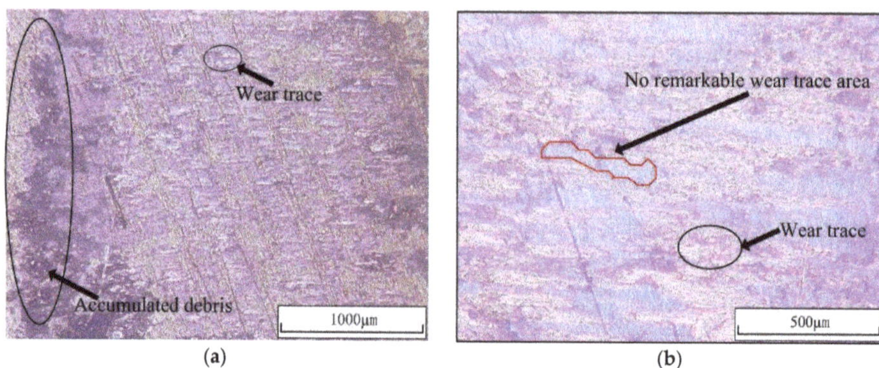

Figure 17. *Cont.*

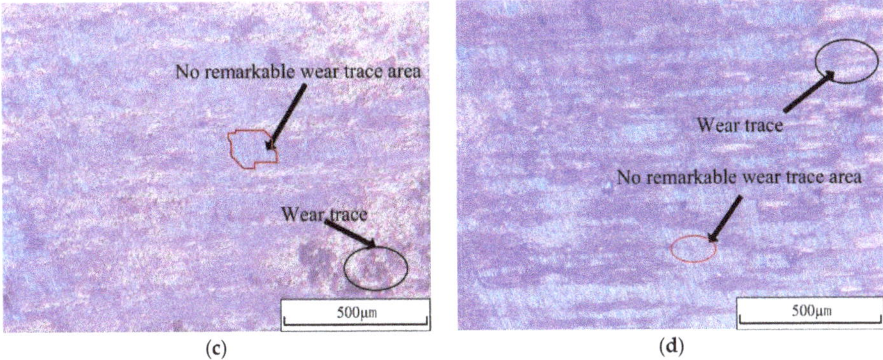

Figure 17. Pictures of the typical wear surfaces of the specimens and pads. (**a**) Wear trace and accumulated debris of untreated specimen. (**b**) Wear trace under laser power density of 3.2 GW/cm^2. (**c**) Wear trace under laser power density of 4.8 GW/cm^2. (**d**) Wear trace under laser power density of 6.4 GW/cm^2.

Figure 18 shows the typical OM micrographs of the fractures of the specimens for the untreated sample and the LSP samples with different laser power densities, and the main crack source is circled in these pictures. It is found that the fatigue crack source of the untreated specimen and the LSP specimen with 3.2 GW/cm^2 laser power density are at the surface of the contact area. However, the fatigue crack source of the LPS specimens with 4.8 GW/cm^2 and 6.4 GW/cm^2 laser power density are at the subsurface or in the interior of the specimens. This phenomenon shows that the surface compressive residual stresses improve the fatigue resistance of the surface, and lead to the transfer of the crack initiation position to the interior of the specimen.

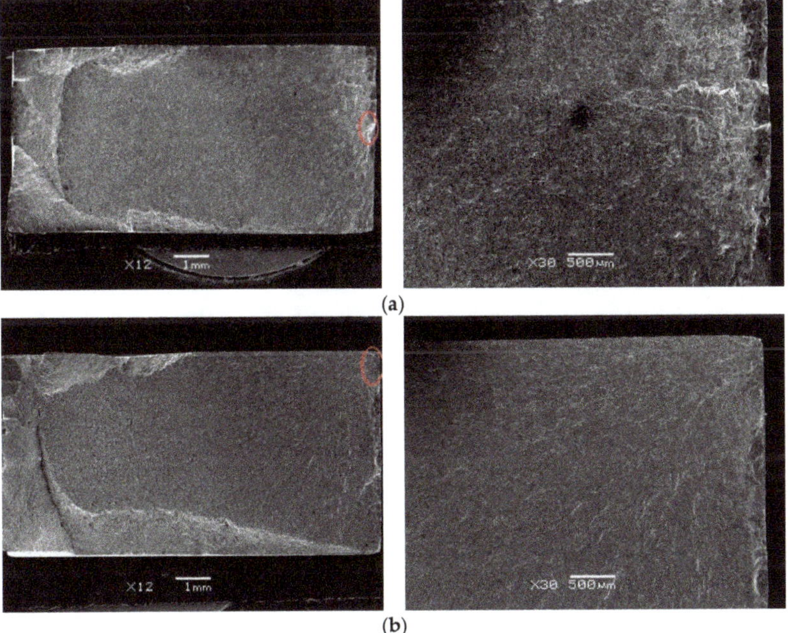

Figure 18. *Cont.*

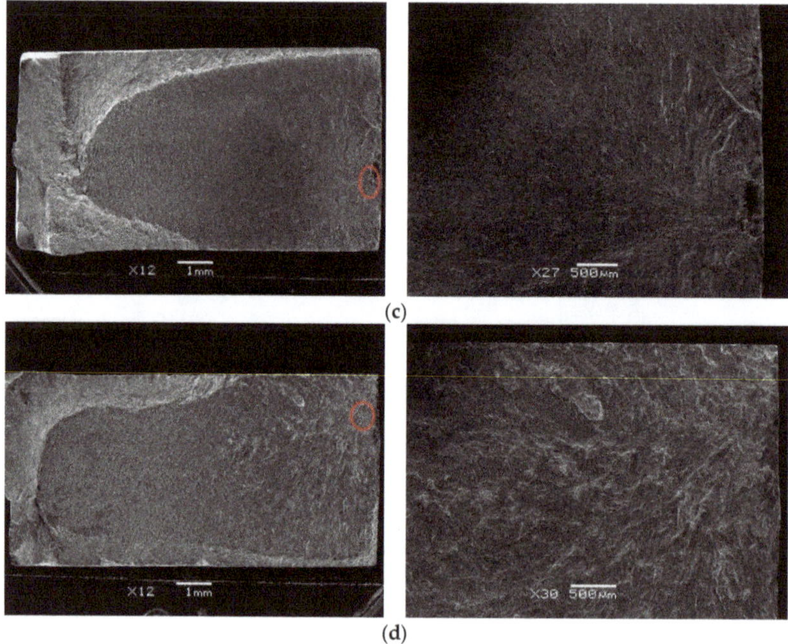

Figure 18. Typical optical microscope (OM) micrographs of the fractures of the specimens. (**a**) Untreated specimen. (**b**) Specimen under laser power density of 3.2 GW/cm^2. (**c**) Specimen under laser power density of 4.8 GW/cm^2. (**d**) Specimen under laser power density of 6.4 GW/cm^2.

The metallographic of a depth of the main specimen, which is near the crack, after laser shock peening (6.4 GW/cm^2) is shown in Figure 19. The formulation of the corrosive agent is as follows: HF:HNO$_3$:H$_2$O = 1:3:15, and the corrosion time is 20 s.

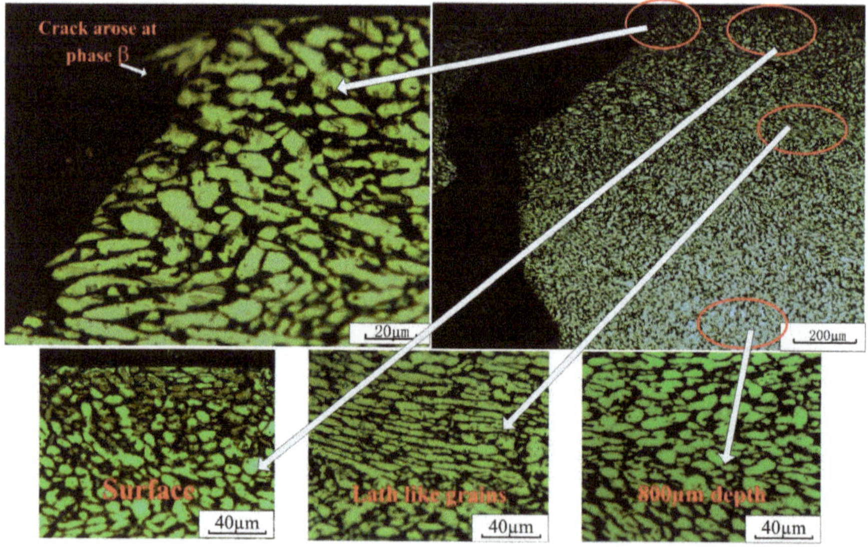

Figure 19. The metallographic of a depth of the main specimen after LSP (6.4 GW/cm^2).

From Figure 19, it can be seen that compared with the grains under the depth of 800 μm, the grain size of the surface layer is smaller due to LSP. The average grain size of the surface layer is about 6.8 μm, while the average grain size of the 800 microns depth is about 13.7 μm. It is also found that due to the effect of LSP, some grains of the specimen are extruded into a lath like state with the grain thickness of about 3.5 μm. The crack arises due to the location of phases β, and passes through the grains (both α and β) in the process of propagation.

4. Conclusions

1. In this paper, the TC11 specimens were under laser shock peening with three different laser power densities of 3.2 GW/cm^2, 4.8 GW/cm^2 and 6.4 GW/cm^2. The surface topography of the untreated and LSP samples were carried out by using a non-contact 3D optical profilometer, and it showed that with the increase of the laser power density, the surface roughness also increased, which meant a worse surface condition.

2. A proto-LXRD X-ray diffractometer was used to measure the surface residual stresses, and the result showed that the laser power density of 4.8 GW/cm^2 had the best effect with the introduction of compressive residual stresses of −650 Mpa, which meant the plastic deformation of the TC11 surface had reached saturation. The excessive increase of the laser power density would cause the peak pressure of the shock wave to be too large, which reduced the surface quality and induced tensile stress on the surface.

3. A specialized fretting pad fixture and fretting fatigue test rig were used to measure the initiation lives of the fretting fatigue crack. With the comparison between untreated and LSP specimens, it was found that the fretting fatigue life of LSP was significantly improved between 2 and 4 times. In addition, the 4.8 GW/cm^2 power density had the best effect on the improvement of fatigue life, although the average surface residual stress was similar at the power density of 4.8 GW/cm^2 and 6.4 GW/cm^2, the high power density caused the bigger surface damage in the material, which led to the reduction of fatigue life.

4. The OM micrographs of the fractures of the specimens showed that with the increase of laser power density, the source of crack initiation was gradually transferred from the surface to interior of the specimen, which meant that the introduction of LSP improved the surface strength and reduced the surface damage of the specimen, and the fretting effect did not occupy the dominant position in the process of crack initiation. The metallographic of the main specimen showed that the crack arose due to the location of phases β, and passed through the grains (both α and β) in the process of propagation.

Author Contributions: Conceptualization, H.Z.; Investigation, X.Y. and C.W.; Project administration, H.Z. and H.C.; Supervision, H.Z. and H.C.; Writing—original draft, X.Y.; Writing—review & editing, H.Z. All authors have read and agreed to the published version of the manuscript.

Funding: This research was funded by National Natural Science Foundation of China, grant number:91860111, National Science and Technology Major Project of China (2017-IV-0012-0049) and Fundamental Research Funds for the Central Universities (No. 1002-XBC19007).

Conflicts of Interest: The authors declare no conflict of interest.

References

1. Hills, D.A. Mechanics of fretting fatigue. *Wear* **1994**, *175*, 107–113. [CrossRef]
2. Golden, P.J.; Hutson, A.; Sundaram, V.; Arps, J.H. Effect of surface treatments on fretting fatigue of Ti–6Al–4V. *Int. J. Fatigue* **2007**, *29*, 1302–1310. [CrossRef]
3. Nicholas, T. Critical issues in high cycle fatigue. *Int. J. Fatigue* **1999**, *21*, 221–231. [CrossRef]
4. Shen, M.; Peng, J.; Zheng, J. Study and Development of Fretting Fatigue. *J. Mater. Eng.* **2010**, *12*, 86–91.

5.   Golden, P.J.; Shepard, M.J. Life prediction of fretting fatigue with advanced surface treatments. *Mater. Sci. Eng. A* **2007**, *468–470*, 15–22. [CrossRef]

6.   Liu, D.; Tang, B.; Zhu, X.; Chen, H.; He, J.; Celis, J.P. Improvement of the fretting fatigue and fretting wear of Ti6Al4V by duplex surface modification. *Surf. Coat. Technol.* **1999**, *116*, 234–238. [CrossRef]

7.   Vantadori, S.; Valeo, J.V.; Zanichelli, A. Fretting fatigue and shot peening: A multiaxial fatigue criterion including residual stress relaxation. *Tribol. Int.* **2020**, *151*, 106537. [CrossRef]

8.   Liu, X.; Liu, J.; Zuo, Z.; Zhang, H. Effects of Shot Peening on Fretting Fatigue Crack Initiation Behavior. *Materials* **2019**, *12*, 743. [CrossRef]

9.   Martín, V.; Vázquez, J.; Navarro, C.; Domínguez, J. Effect of shot peening residual stresses and surface roughness on fretting fatigue strength of Al 7075-T651. *Tribol. Int.* **2020**, *142*, 106004. [CrossRef]

10.   Zhang, H.; Yang, X.; Cui, H.; Wen, W. Study on the Effect of Laser Quenching on Fretting Fatigue Life. *Metals* **2019**, *9*, 566. [CrossRef]

11.   Srinivasan, S.; Garcia, D.B.; Gean, M.C.; Murthy, H.; Farris, T.N. Fretting fatigue of laser shock peened Ti–6Al–4V. *Tribol. Int.* **2009**, *42*, 1324–1329. [CrossRef]

12.   Nie, X.; He, W.; Zang, S.; Wang, X.; Zhao, J. Effect study and application to improve high cycle fatigue resistance of TC11 titanium alloy by laser shock peening with multiple impacts. *Surf. Coat. Technol.* **2014**, *253*, 68–75. [CrossRef]

13.   Tong, Z.P.; Ren, X.D.; Zhou, W.F.; Adu-Gyamfi, S.; Chen, L.; Ye, Y.X.; Ren, Y.P.; Dai, F.Z.; Yang, J.D.; Li, L. Effect of laser shock peening on wear behaviors of TC11 alloy at elevated temperature. *Opt. Laser Technol.* **2019**, *109*, 139–148. [CrossRef]

14.   Qiao, H.; Zhao, J. Analysis and Optimization on Laser Peening Parameters of Tianium Alloy. *Acta Optiac Sinaca* **2013**, *33*. [CrossRef]

15.   Lei, Z.Z.; Yinghong, L.L.; Cheng, W.W.; Xin, Z. Laser shock peening for LY2 alloy. *High. Power Laser Part. Beams* **2010**, *22*, 1780–1784.

16.   Ren, X.D.; Zhang, Y.K.; Zhou, J.Z.; Kong, D.J. Thickness optimizing of surface coating for laser shock processing. *Heat Treat. Met.* **2006**, *31*, 61–64.

17.   Lee, H.; Sathish, S.; Mall, S. Evolution of residual stress under fretting fatigue. *J. Mater. Sci.* **2004**, *39*, 7089–7092. [CrossRef]

18.   King, A.; Steuwer, A.; Woodward, C.; Withers, P.J. Effects of fatigue and fretting on residual stresses introduced by laser shock peening. *Mater. Sci. Eng. A* **2006**, *435*, 12–18. [CrossRef]

19.   Liu, K.K.; Hill, M.R. The effects of laser peening and shot peening on fretting fatigue in Ti–6Al–4V coupons. *Tribol. Int.* **2009**, *42*, 1205–1262. [CrossRef]

20.   Qiao, H.; Hu, X.; Zhao, J.; Wu, J.; Sun, B.; Lu, Y.; Guo, Y. Influence Parameters and Development Application of Laser Shock Processing. *Surf. Technol.* **2019**, *48*, 1–9.

21.   Fabbro, R.; Fournier, J.; Ballard, P.; Devaux, D.; Virmont, J. Physical study of laser-produced plasma in confined geometry. *J. Appl. Phys.* **1990**, *68*, 775–784. [CrossRef]

22.   Fabbro, R.; Peyre, P.; Berthe, L.; Scherpereel, X. Physics and applications of laser-shock processing. *J. Laser Appl.* **1998**, *10*, 265. [CrossRef]

23.   Deng, Z.H.; Liu, Q.B.; Xu, P.; Yao, Z.H. Corrosion resistance and mechanism of metallic surface processed by square-spot laser shock peening. *J. Mater. Eng.* **2018**, *46*, 140–147.

24.   Jiang, Y.F.; Ji, B.; Gan, X.D.; Hua, C.; Li, X.; Zhu, H. Study on the effect of laser peening with different power densities on fatigue life of fastener hole. *Opt. Laser Technol.* **2018**, *106*, 311–320. [CrossRef]

25.   Hu, Y.X. Research on the Numerical Simulation and Impact Effects of Laser Shock Processing. Ph.D. Thesis, Shanghai Jiaotong University, Shanghai, China, 2008.

26.   Peyre, P.; Fabbro, R.; Merrien, P.; Lieurade, H.P. Laser shock processing of aluminium alloys. Application to high cycle fatigue behaviour. *Mater. Sci. Eng. A* **1996**, *210*, 102–113. [CrossRef]

27.   Li, B.M.; Liu, X.M.; Zhang, H.H.; Liu, F. Numerical simulation of laser shock processing in 2024 aluminum alloy. *Appl. Laser* **2017**, *37*, 852–858.

28.   Hua, G.; Jiang, S.; Cao, Y.; Zhou, D.; Chen, H. Numerical simulation of residual stress hole on 7050 aluminum alloy under laser shock. *Heat Treat. Met.* **2017**, *42*, 154–157.

29. Sun, C.W. *The Effect of Laser Irradiation*; National Defence Industry Press: Beijing, China, 2002.
30. Aeronautical Manufacture Engineering Committee. *Aeronautical Manufacture Engineering Handbook*; Aerospace Industry Press: Beijing, China, 1997.

Publisher's Note: MDPI stays neutral with regard to jurisdictional claims in published maps and institutional affiliations.

Article

The Effect of Laser Re-Solidification on Microstructure and Photo-Electrochemical Properties of Fe-Decorated TiO$_2$ Nanotubes

Piotr Kupracz [1,*], **Katarzyna Grochowska** [1,*], **Jakub Karczewski** [2], **Jakub Wawrzyniak** [1] **and Katarzyna Siuzdak** [1]

[1] Centre of Plasma and Laser Engineering, The Szewalski Institute of Fluid-Flow Machinery PASci, Fiszera 14 Street, 80-231 Gdańsk, Poland; jwawrzyniak@imp.gda.pl (J.W.); ksiuzdak@imp.gda.pl (K.S.)

[2] Faculty of Applied Physics and Mathematics, Gdańsk University of Technology, Narutowicza 11/12 Street, 80-233 Gdańsk, Poland; jakub.karczewski@pg.edu.pl

* Correspondence: pkupracz@imp.gda.pl (P.K.); kgrochowska@imp.gda.pl (K.G.)

Received: 23 July 2020; Accepted: 8 September 2020; Published: 10 September 2020

Abstract: Fossil fuels became increasingly unpleasant energy source due to their negative impact on the environment; thus, attractiveness of renewable, and especially solar energy, is growing worldwide. Among others, the research is focused on smart combination of simple compounds towards formation of the photoactive materials. Following that, our work concerns the optimized manipulation of laser light coupled with the iron sputtering to transform titania that is mostly UV-active, as well as exhibiting poor oxygen evolution reaction to the material responding to solar light, and that can be further used in water splitting process. The preparation route of the material was based on anodization providing well organized system of nanotubes, while magnetron sputtering ensures formation of thin iron films. The last step covering pulsed laser treatment of 355 nm wavelength significantly changes the material morphology and structure, inducing partial melting and formation of oxygen vacancies in the elementary cell. Depending on the applied fluence, anatase, rutile, and hematite phases were recognized in the final product. The formation of a re-solidified layer on the surface of the nanotubes, in which thickness depends on the laser fluence, was shown by microstructure studies. Although a drastic decrement of light absorption was recorded especially in UV range, laser-annealed samples have shown activity under visible light even 20 times higher than bare titania. Electrochemical analysis has shown that the improvement of photoresponse originates mainly from over an order of magnitude higher charge carrier density as revealed by Mott-Schottky analysis. The results show that intense laser light can modulate the semiconductor properties significantly and can be considered as a promising tool towards activation of initially inactive material for the visible light harvesting.

Keywords: TiO$_2$ nanotubes; Fe$_2$O$_3$; iron oxide nanoparticles; water splitting; laser processing

1. Introduction

The increasing global energy from renewable sources demand of well-developed society conflicts with the structure of the world energy balance, where fossil fuels take over 80% [1]. The most promising candidate for the modern energy source is solar energy, as it can be easily converted into electric power almost everywhere. However, the fluctuant insolation and limited electric energy storage technology make its application unsuitable for long-range transport and long-term storage. Photo-generation of H$_2$ from water seems to be one of very attractive technologies of solar power accumulation due to easiness of chemical to electrical energy conversion [2]. Unfortunately, the efficiency of the solar enhanced water splitting systems suffers from the lack of low-cost and highly effective electrode materials. Currently, two main material classes are considered to be the most suitable namely nanoscale noble metals and semiconductors. Since the former is very expensive, the researchers focus on photoactive semiconductors [3]. For example, the perfect photo-anode devoted to the water splitting system should have direct optical bandgap around 2.2 eV, low overpotential in reference to oxygen or hydrogen evolution reaction, and low recombination rate of charge carriers [4,5]. In the case of the photoactive anode material, semiconductors like RuO_2, Pb_3O_4, and CdS are the nearest to satisfying these requirements; however, CdS undergoes corrosion during exposition to light, and utilization of Pb and Cd is limited due to their toxicity, while Ru is very expensive [6]. On the other hand, ZnO_2, TiO_2, $BaTiO_2$, and SnO_2 are generally regarded as non-toxic, chemically inert, and inexpensive but exhibit higher electron-hole recombination rate and much wider bandgap (3.0–3.5 eV) than RuO_2, Pb_3O_4, and CdS [7,8], which limits their effective light-gathering ability to less than 6% of solar spectrum power. Nevertheless, it should be realized that free zinc oxide or titania nanoparticles can negatively impact the DNA chains and proteins [9,10], but formation of the material already fixed onto the stable substrate and their further usage in the immobilized form minimizes this risk.

Fujishima and Honda [11], as a first, focused world attention on the photo-electrochemical properties of TiO_2 building the first solar-driven electrochemical cell. However, works on its application in the water-splitting system have accelerated after the successful dye sensitization of TiO_2, which extends light absorption to the visible part of the spectrum [12]. These two seminal works became the basis for the research on a wide branch of modified TiO_2 for electrodes in water-splitting systems.

Since the recombination rate of photo-generated minority carriers depends on the diffusion path length to the electrode surface, where they can take part in a water molecule oxidation [13], the structural engineering is especially focused in a way to produce highly porous structures. As it has been already shown [14–20], the preparation of TiO_2 in a form of various nanostructures, like nanobelts, nanorods or nanotubes drastically improves photo-electrodes performance in a reference to the plane material. However, its tubular structure synthesized by anodization is most frequently examined due to perpendicular orientation of hollow cylinders regarding the metal substrate reducing the charge recombination and the scalability of the fabrication process [15,21]. Additionally, the geometrical parameters of titania nanotubes (TiO_2-NTs) could be easily optimized during the synthesis, giving rise to the specific surface and light absorption of the resulting material [14,22]. The advantage of the direct contact between the electrode material and the current collector, as well as well-developed surface area, has been used in prototype supercapacitors when the further hydrogenation via cathodic polarization or plasma treatment in hydrogen atmosphere is applied. Further improvement of TiO_2-NTs electrochemical performance could be achieved by the formation of junction between TiO_2 and covering semiconductors with higher positioned valence and conduction bands which promotes dissociation of photo-generated electron-hole pair [23]. Such an approach has more advantages since the presence of well-adjusted semiconductor on TiO_2-NTs can decrease the Shottky barrier height at the electrode-electrolyte interface [6,15] and extends the range of the heterojunction photoresponse from ultraviolet (UV) to the visible region. Recently, the usage of transition metal oxides, like NiO, Fe_2O_3, or Cu_2O [6,24–27], has been extensively explored as the decoration can be achieved via various methods, like chemical deposition [28], electrochemical deposition [29], alternative layer deposition [30], hydrothermal deposition [23], or photodecomposition [31]. Among others, the TiO_2-α-Fe_2O_3 heterojunction is the most explored due to the well-suited for light gathering bandgap (2.2 eV) of the latter, its low cost, and a relative misalignment of the semiconductors band edges promoting holes and electrons accumulation in TiO_2 and α-Fe_2O_3, respectively [24,32–34]. Further junction enhancement is observed when pseudobrookite (TiO_5) is formed at the interface between hematite and titania [34–36]. Due to its narrower bandgap (2.1 eV) and alignment of valence and conduction bands, the TiO_2-Fe_2TiO_5-α-Fe_2O_3 heterojunction additionally facilitates solar absorption and promotes the charge carriers' separation.

Apart from the modification by other metal oxides, another strategies to improve the performance of pristine TiO_2 based electrode cover, e.g., treatment with plasma, annealing in hydrogen, and laser irradiation [37]. Among others, the ns laser pulse induces fast heating and cooling process, even above the melting point, therefore, its interaction with multi-layer composite may result in diffusion of atoms across the interphase, as well as the formation of metastable phases and structural defects [38]. That technique is extremely fast, is easy-scalable for large areas, and due to a wide range of processing parameters can be applied to various materials [39]. The modification of anatase TiO_2-NTs electrode using Nd:YAG laser was shown to enhance charge carrier donor density over one order of magnitude and a shift of the flat-band potential up to +0.7 V vs. Ag|AgCl|0.1 M KCl [16,39]. However, the modification leads to the photo-response improvement by no more than a few tens of percent. On the other hand, doping TiO_2 by Fe decreases recombination of the photogenerated charge carriers, and provides narrower bandgap [40–42], while charge carriers density in Ti-doped hematite is one order of magnitude higher in reference to the pristine [35]. Although the coverage of TiO_2 by Fe_2O_3, and doping TiO_2 with Fe, as well as Fe_2O_3 with Ti, and laser treatment of TiO_2-NTs lead to the improvement on the electrode performance, the combined approach has not been examined till now.

In this work, an array of parallel TiO_2-NTs were electrochemically fabricated and covered by several nm thick Fe layer. As obtained hetero-structures were further modified by the pulsed Nd:YAG laser treatment, followed by the calcination in an electrical furnace. The main attention was paid to the impact of metal layer thickness, and laser fluence on the electrochemical performance of Fe-decorated TiO_2-NTs samples. Surface morphology and composition analyses were performed using scanning electron microscopy, Raman spectroscopy, X-ray diffraction, and X-ray photoelectron spectroscopy. The diffuse reflectance spectroscopy was employed to study the light absorption capability and optical band-gap position. Electrochemical measurements have proven that easily scalable laser processing over the titania NTs with sputtered thin iron film can significantly change the electric properties, namely resulting in the cathodic shift of the flat-band potential and growth of negative charge carriers density, which inflicts 20-times enhancement of the charge carriers photo-generation efficiency under visible light irradiation compared to the pristine TiO_2-NTs.

2. Materials and Methods

Highly-oriented titanium nanotubes arrays were produced by one-step electrochemical anodization in a two-electrode arrangement. The titanium foil (99.7%, Strem) with dimensions of 2×3 cm^2 acted as an anode and Pt rectangular mesh as a cathode in the electrolyte consisted of 0.27 M NH_4F (p.a. Chempur) and 15 wt% of deionized water in ethylene glycol (p.a. Chempur), similar to other treatment procedures [43]. The electrolyte temperature was kept at 23 °C with the thermostat (Julabo F-12, Julabo, Seelbach, Germany). The experimental setup used for anodization is shown in Figure 1a. Ti foil was cleaned by a three-step process, i.e., firstly the 10-min ultrasonic bath in acetone (Protolab, Słupsk, Poland), next in ethanol (96%, Chempur, Piekary Śląskie, Poland), and finally in deionized water (0.08 μS, Hydrolab HLP-5p, Hydrolab, Straszyn, Poland). Acetone and ethanol are used for degreasing and removing of the organic contamination while final water treatment simply removes the residues of acetone and ethanol because in the main anodization bath those compounds are not present. After all, samples were rinsed with isopropanol (p.a. POCH) and left to dry in the air. A detailed description of the anodization process can be found in Reference [26]. The titanium foil with as-growth TiO_2-NTs was annealed in an electrical furnace (Nabertherm, Lilienthal, Germany) in the static air with a heating rate of 2 °C/min up to 450 °C, kept 2 h, and cooled freely. Next, TiO_2-NTs serve as a support for iron deposition utilizing magnetron sputtering (Q150S, Quorum Technologies, Lewes, UK) with mounted highly pure Fe target (99.5 % EM-Tec). The Fe thickness was controlled by quartz microbalance and implemented program and set to 5, 10, and 15 nm.

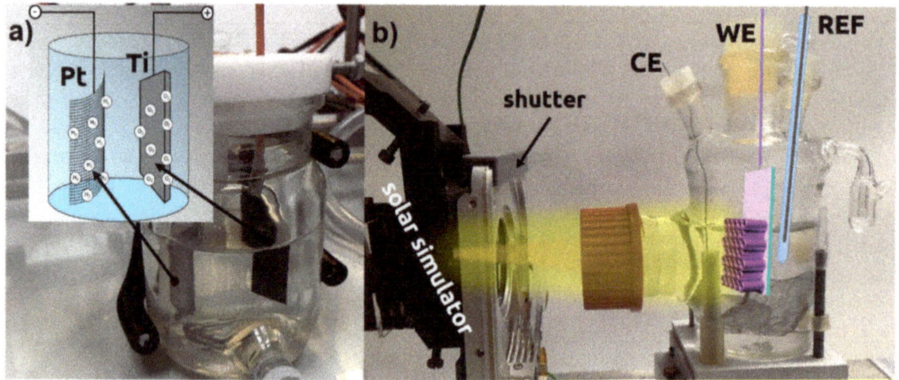

Figure 1. The experimental setups used for (a) the anodization and (b) photoelectrochemical characterization.

Next, the pulsed laser modification of Fe-decorated TiO_2-NTs was carried out using a Q-switched Nd:YAG (Quantel, Lannion, France) pulse laser generating 6 ns pulses at the wavelength of 355 nm with the repetition rate of 2 Hz. The laser beam went through a square homogenizer and 200-mm focal length lens giving a 2.7-mm wide square spot of uniform irradiance. In order to increase the uniformity and size of the modified area to ~30 mm^2, the samples were moved back and forth in accordance to laser beam with a constant speed of 3.7 mm/min during the modification. The surface of samples was laser-annealed with an energy fluence (F) of 40, 80, 120, 160, 200 mJ/cm^2 under the vacuum of 3×10^{-5} mbar. After all, samples calcination was repeated maintaining the same parameters. As obtained samples are further referred as FeXLY, where X is the Fe thickness, and Y is the laser fluence applied during the annealing.

The samples' morphology was examined with Scanning Electron Microscope FE-SEM, FEI Quanta FEG 250 (Thermo Fisher, Waltham, MA, USA) with a secondary electron detector maintaining 10 kV accelerating voltage. The X-Ray Photoelectron Spectroscopy studies were carried out in the Ti2p, Fe2p, O1s, and C1s binding energy range using Escalab 250Xi spectroscope (Thermo Fisher, Waltham, MA, USA). The spectroscope was equipped in Al Kα monochromatic X-ray source (spot diameter 650 µm), operating at 20 eV pass energy. Charge compensation was provided utilizing a flood gun, with the final calibration of the obtained spectra based on the signal derived from adventitious carbon (C1s at 284.6 eV) [44]. The structure analysis was carried out by X-ray diffractometry (XRD) over the range of 20°–90° with a Bruker D2Phaser (Bruker, Billerica, MA, USA) diffractometer with CuKα radiation, equipped with a XE-T detector, and the Raman spectroscopy using a confocal micro-Raman spectrometer (InVia, Renishaw, Wotton-under-Edge, Gloucestershire, UK) with an argon-ion laser source emitting at 514 nm and operating with the power of 5 mW. Reflectance spectra were taken by diffusive reflectance spectroscopy with PerkinElmer Lambda 35 (PerkinElmer, Waltham, MA, USA) dual-beam spectrophotometer in the range of 200–900 nm with a scanning speed of 120 nm/min.

The electrochemical properties were measured in a three-electrode system with Autolab PGStat 302 N potentiostat-galvanostat Metrohm (Autolab, Utrecht, Holland) system, where Pt mesh and Ag|AgCl|0.1 M KCl served as a counter (CE) and the reference electrode (REF), respectively. The investigated material was used as working electrode (WE). The 0.5 M NaOH (p.a. POCH) electrolyte was initially deaerated with argon (5N, Air Liquide) and separated from oxygen by Ar continuous flow. The sun-simulator built of 150 W Xenon lamp (Optel, Opole, Poland) equipped with the AM 1.5 filter and additional mobile GG420 UV filter (Schott, Mainz, Germany) was used to perform photo-electrochemical measurements under the light of full solar light and limited to visible spectra, respectively. The whole experimental setup dedicated for photoresponse investigations is presented in Figure 1b.

Cyclic voltammetry (CV), as well as linear voltammetry (LV) measurements, were carried out in the potential range from −1.1 V to +0.9 V vs. Ag|AgCl|0.1 M KCl, with the scan rate of 50 mV/s and 10 mV/s, respectively. The chronoamperometry (CA) under varying light conditions was performed with the working electrode polarization set to +0.4 V vs. Ag|AgCl|0.1 M KCl. The electrochemical impedance for the Mott-Schottky analysis was measured at a single frequency of 1 kHz with a 10-mV amplitude of the sinusoidal signal and in the potential range from −1.0 to +0.8 V versus Ag|AgCl|0.1 M KCl. The stabilization time of 30 s before each impedance record was applied to reduce the influence of charging/discharging processes. The flat-band potential (E_{FB}) and majority charge carriers density (N_d) were calculated according to the Mott-Schottky Equation (1) [45]:

$$\frac{1}{C_{SC}^2} = \frac{2}{\varepsilon_0 \varepsilon_R A e N_d}(E_{FB} + E + k_B T), \tag{1}$$

where C_{SC} is the capacitance of the space charge layer, E is bias potential, ε_0 and ε_R are the vacuum and TiO$_2$ permittivity, and e stays for the elementary charge. For the calculations, the following values were taken: $\varepsilon_R = 38$ [46] for TiO$_2$ (anatase), $\varepsilon_0 = 8.85 \times 10^{-12}$ F/m, and $e = 1.602 \times 10^{-19}$ C.

3. Results and Discussion

The microstructure of TiO$_2$-NTs covered by 10 nm of Fe and samples after the laser treatment of 355 nm wavelength is shown in Figure 2. The TiO$_2$-NTs height, diameter, wall thickness, and distance between nanotubes were measured to be ca. 1500 nm, 117 nm, 17 nm, and 150 nm, respectively. The laser treatment with a fluence as low as 40 mJ/cm^2 (Figure 2b) leads to the partial transformation of the TiO$_2$-NTs surface to the 160 nm thick porous layer and in consequence to nanotubes height reduction (Figure 3). As the higher fluence of laser beam was applied (Figure 2c–f), a stronger shortening is observed until almost complete degradation of the well-ordered titania construction occurs for F = 200 mJ/cm^2. The re-solidified surface layer porosity decrement accompanies that phenomenon, and its thickness grows from 160 nm to 405 nm as laser fluence increases from 40 mJ/cm^2 to 200 mJ/cm^2.

The effect of laser annealing on the pristine TiO$_2$-NTs was studied by several authors [16,17,37,39,47]. In all cases, it was concluded that the laser fluence required to create the continuous re-solidified layer has to be higher than 40 mJ/cm^2. Additionally, they have shown that, using that fluence, the TiO$_2$-NTs rather join with the neighbors, creating stuck islands, than forming the compact thick layer. Comparing their results with the morphology of Fe-decorated TiO$_2$-NTs, one may conclude that the presence of Fe plays an important role in the re-solidifying process. The absorption coefficients of Fe, Fe$_2$O$_3$, Fe$_3$O$_4$, and TiO$_2$ for light of 355 nm wavelength are 1.15×10^6 cm^{-1} [48], 3.95×10^5 cm^{-1}, 3.95×10^4 cm^{-1} [49], and 4.12×10^3 cm^{-1} [50], respectively. Since iron and iron oxides absorb the incident laser orders of magnitude stronger than TiO$_2$, it becomes clear that the same laser fluence induces more efficient heating at the surface of Fe-decorated TiO$_2$-NTs than bare ones leading to its more uniform re-solidification; please see work of Haryński et al. [39] for reference.

The dependency between TiO$_2$-NTs length and the re-solidified layer is shown in Figure 3. As could be seen, the height of TiO$_2$-NTs and the thickness of the re-solidified layer change almost linearly with the laser fluence from 1500 nm to 550 nm and from 0 nm to 410 nm, respectively. Taking into account their closed packed microstructure, their effective volume fill factor can be calculated by Equation (2):

$$V_{\text{fill}} = \frac{a^2 \sqrt{3}}{2\pi(R^2 - r^2)}, \tag{2}$$

where a, R, and r, are separation, external and internal diameter of TiO$_2$-NTs, respectively. Substituting the dimensions of bare TiO$_2$-NTs from Figure 2, V_{fill} equals to ca. 0.45. Since the re-solidified layer in SEM images seems to be almost completely devoid of bubbles, the total volume of the forming layer should be nearly two times lower than unmodified TiO$_2$-NTs. That explanation perfectly matches the registered evolution of TiO$_2$-NTs morphology under the laser treatment (Figure 3a). On the other hand, the increasing thickness of the iron film on the sample treated with 80 mJ/cm^2 fluence causes less intensive TiO$_2$-NTs reduction under laser irradiation, while the measured thickness of the re-solidified layer remains almost constant (Figure 3b). It means that the porosity of re-solidified layer depends on the amount of Fe onto titania support. As shown above, stronger laser fluence leads to the less porous re-solidified layer. Therefore, the higher amount of Fe, the weaker heating takes place. The reflectivity measured for bulk Fe, Fe$_2$O$_3$, Fe$_3$O$_4$, and TiO$_2$ for light of 355 nm wavelength is 0.68 [48], 0.25, 0.17 [49], and 0.25 [50], respectively. Thus, presence of pure iron may reduce the effectiveness of the laser annealing. It leads to the conclusion that the Fe-layer may play an opposite role. The small amount of Fe could increase the light absorption, leading to the uniform melting of the surface, while the thicker layer reflects the 355 nm radiation, reducing the effect of the laser annealing on the microstructure.

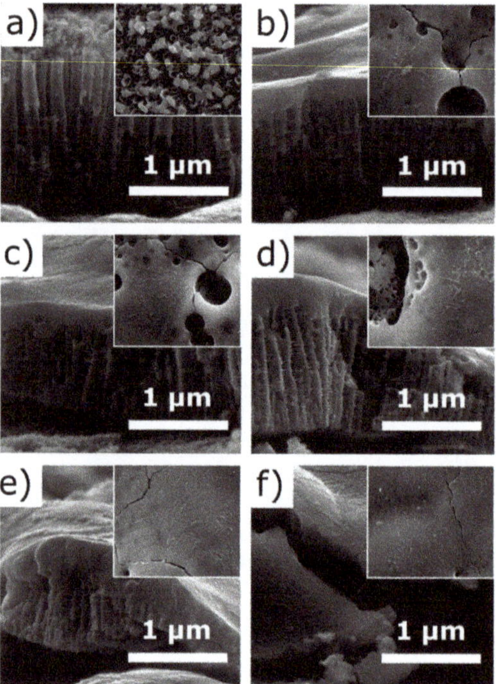

Figure 2. SEM images of laser-treated titania nanotubes (TiO$_2$-NTs) covered by 10 nm Fe film from top (inset) and side. Images show the samples (**a**) untreated and treated by laser of (**b**) 40 mJ/cm^2, (**c**) 80 mJ/cm^2, (**d**) 120 mJ/cm^2, (**e**) 160 mJ/cm^2, and (**f**) 200 mJ/cm^2 fluence.

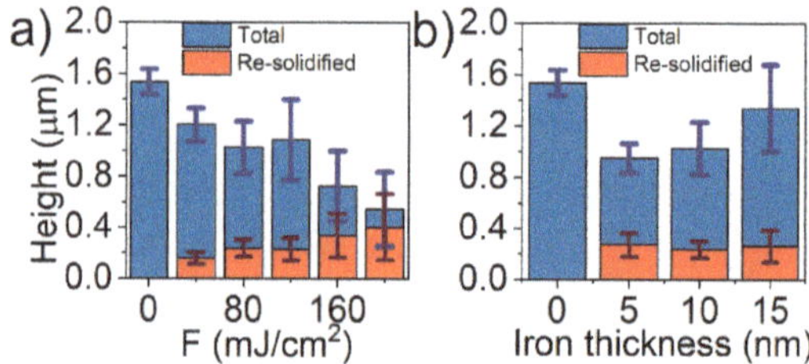

Figure 3. The dependence of the length of titania nanotubes (**a**) with 10 nm deposited Fe on the applied laser fluence and (**b**) on the iron layer thickness when the applied laser fluence equals 80 mJ/cm^2.

X-ray diffraction patterns of modified TiO_2-NTs are shown in Figure 4. Three major phases were recognized in the pattern, namely metallic Ti substrate with assigned peaks at 34.8°, 38°, 40°, 53°, 63° and 70° (PDF-2 ICDD: 00-044-1294), anatase TiO_2: 25°, 37.8°, 48°, 53.7°, and 55° (96-900-8214), and rutile TiO_2: 27°, 36°, 41°, and 69° (96-900-9084). Any other peaks which may reflect the presence of Fe, FeO, Fe_3O_4, and Fe_2O_3 are not seen. It is worthy of note that the shift towards higher angles is observed for the position of the peak at 25° (inset Figure 4), and, for the Fe10L40, Fe5L80, Fe10L80, Fe15L80, Fe10L120, and Fe10L160 samples, $\Delta 2\Theta$ is 0.05°, 0.08°, 0.05°, 0.04°, 0.21°, and 0.19°, respectively. As seen, the shift induced by the laser annealing increases with the fluence and reaches the maximum for 120 mJ/cm^2. On the other hand, growing iron thickness reduces the structural changes induced by the treatment. In general, a positive shift of the diffraction pattern originates from the shrinkage of a primitive cell. Since TiO_2-NTs were annealed in a presence of iron in a vacuum, it is possible that crystallographic cell shrinkage is induced by the oxygen escape from TiO_2 lattice with an accompanying reduction of Ti^{4+} to Ti^{3+} [51] or by an additional substitution of Ti^{3+} by smaller Fe^{3+} ions [42].

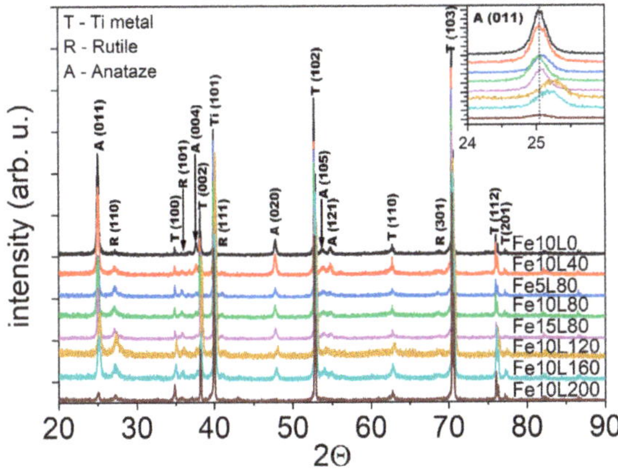

Figure 4. X-ray diffraction patterns of laser-annealed Fe-decorated TiO_2-NTs.

The crystal structure of the laser-annealed Fe-decorated TiO_2-NTs samples was distinguished by the Raman spectroscopy. As shown in Figure 5, five main peaks at 144, 196, 395, 515, and 634 cm^{-1} can be assigned to anatase $E_{g(1)}$, $E_{g(2)}$, $B_{g(1)}$, $A_{g(1)}$, and $E_{g(3)}$ phonon modes, respectively [52]. However, with an increasing laser fluence from 40 to 200 mJ/cm^2, the relative intensity of the anatase signal drops and peaks at 447 and 612 cm^{-1} originating from rutile $E_{g(1)}$ and $A_{g(1)}$ modes [53] become observable. Additionally, weak peaks at 1320 and 1576 cm^{-1} are seen (Figure 5c). The former arises from the interaction of two magnons created on antiparallel close spin sites [54] or from a second-order tone of a forbidden phonon mode [55], while the latter is ascribed to free magnon scattering [55]. Raman bands for hematite and magnetite expected at 550 and 660 cm^{-1} and at 193, 306, 538, and 668 cm^{-1}, respectively [54,56], were not recorded due to their overlapping with stronger bands originated from anatase and rutile.

The variation of the relative intensity of the α-Fe_2O_3 and rutile-TiO_2 compared to the anatase-TiO_2 peaks imply that the laser annealing reduces the amount of the crystalline phase. Additionally, signal positions corresponding to main $E_{g(1)}$ active anatase mode of laser-annealed samples depend on the utilized energy fluence during laser modifications; $E_{g(1)}$ is located at 144.4, 143.8, 146.5, 146.5, 145.8 and 144.9 cm^{-1} for 0, 40, 80, 120, 160, and 200 mJ/cm^2 energy fluence, respectively (Figure 5b). A blue shift of Raman peaks is usually referred to the presence of oxygen vacancies originating from distortion of crystal structure in defective TiO_2 [57] or by substitution of Ti^{4+} by Fe^{3+} ions [58]. Thus, samples treated with 80 mJ/cm^2 and 120 mJ/cm^2 are expected to have the highest oxygen vacancies concentration. The strongest positive Raman and XRD shifts were observed for the

Fe10L80 sample. In order to examine the influence of the iron layer thickness on the modified material structure, two samples, namely Fe5L80 and Fe15L80, were studied. However, Raman spectroscopy, as well as XRD pattern, have shown that Fe10L80 undergoes stronger structural evolution due to laser annealing; thus, further examinations have been limited to the sample covered by 10 nm of Fe.

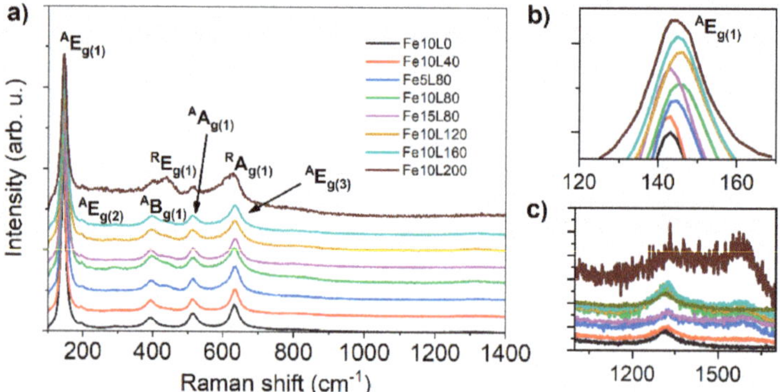

Figure 5. (a) Raman shift spectra recorded for laser-modified Fe-decorated TiO_2-NTs. (b) The magnification of the anatase $E_{g(1)}$. (c) Magnon scattering peaks.

The high-resolution XPS analysis allows providing insight into the surface chemistry of the investigated Fe80L10 sample. In Figure 6, the survey spectrum of the material is presented, where signals attributed to titanium, oxygen, iron, and carbon are indicated. Following that the high-resolution spectra were recorded and analyzed.

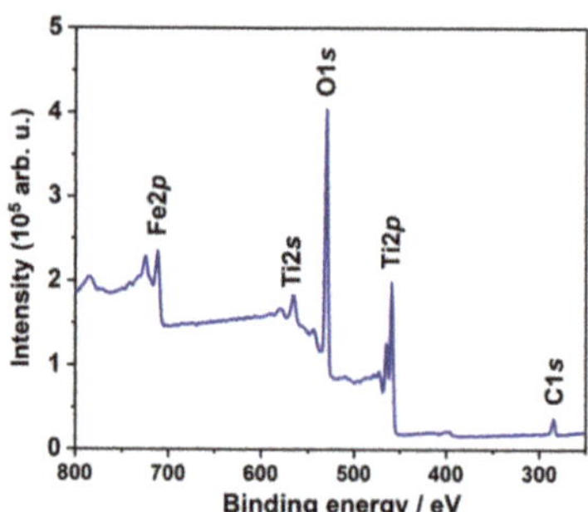

Figure 6. The XPS survey spectrum recorded for Fe10L80 sample.

The Ti2p spectra are composed of a single chemical state, represented by the peak doublet where Ti2p$_{3/2}$ is positioned at 458.8 eV (Figure 7a). The peaks in the deconvoluted energy range are characteristic for the TiO_2 nanotubes as reported in our previous studies [39,59]. Although XRD and Raman spectroscopy may suggest the presence of Ti^{3+} ions, no other oxidation states were present confirming that the chemistry of TiO_2-NTs was not significantly altered during the modification process. It supports the previous assumption that the structural changes may be caused by substitution of Ti^{4+} by Fe^{3+} ions. On the other hand, the broad spectra recorded in the Fe2p energy range is peaking at 710.9 eV (Figure 7b). The position of Fe2p$_{3/2}$ peak, as well as complex multiplet-split peak shape, reveals its origin as Fe_2O_3. This observation is further confirmed by Fe^{3+} satellite visible at approx. 720 eV [60,61]. These findings are in good agreement with the nature of O1s components (see Figure 7c), two of which were identified and deconvoluted within this model. The primary component, at 530.1 eV, is characteristic for both TiO_2 and Fe_2O_3. The second and smaller component could be attributed to surface hydroxyl species but also adventitious carbon contamination [39,44]. According to the surface area, the atomic share of particular elements has been determined: Ti: 18.3%, C: 11.3 %, O: 66.8%, and Fe: 3.6%.

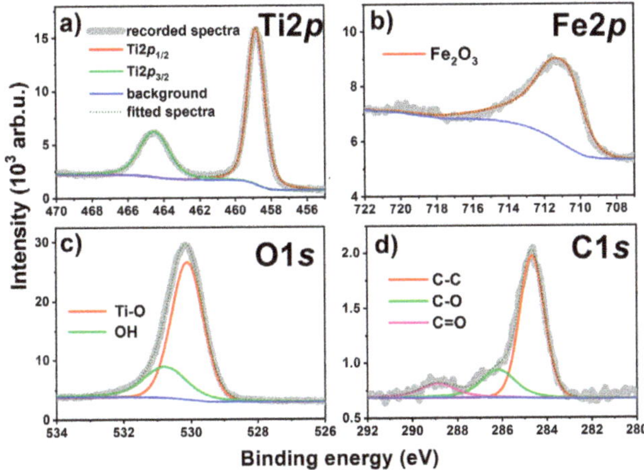

Figure 7. The XPS high resolution spectra registered for Fe10L80 sample: (**a**) titanium, (**b**) iron, (**c**) oxygen, and (**d**) carbon.

The optical properties of samples were examined by diffuse reflectance spectroscopy. The Kubelka-Munk function (F(R)) as a wavelength function is shown in Figure 8. As seen, the laser treatment with the fluence of 40 mJ/cm^2 and higher drastically reduced absorption in the UV range. As a consequence, the determination of optical bandgap (BG) becomes possible only for Fe10L0 and Fe0L0 samples. Their values were determined based on the Tauc plot for indirect allowed transition as 1.82 and 2.92 eV, respectively. The pristine anatase, rutile, hematite, and magnetite are semiconductors with the bandgap values of 3.2, 3.0, 2.2, and 0.1 eV, respectively [62,63], while phases composed of Fe_2O_3-TiO_2 [64–66] are expected to exhibit BG in the range of 1.5–1.9 eV. Therefore, the observed optical bandgap of Fe10L0 sample can be ascribed to the presence of non-stoichiometric or the doped hematite.

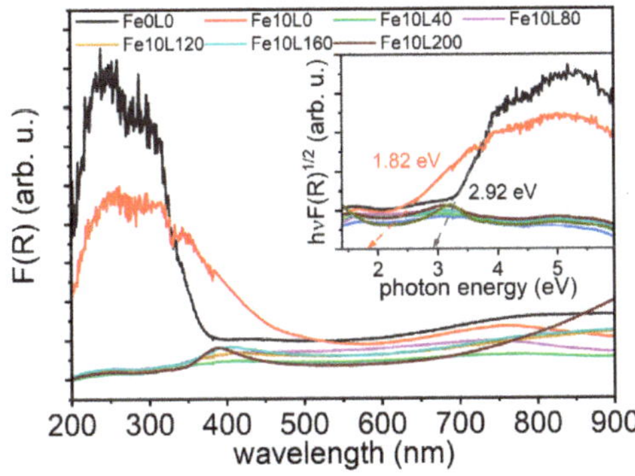

Figure 8. Absorbance spectra and Tauc plots (inset) for laser-treated Fe-decorated TiO_2-NTs.

Cyclic voltammetry scans of the unmodified and laser-annealed Fe-decorated TiO_2-NTs with the fluence of 80 mJ/cm^2 are shown in Figure 9. All TiO_2-NTs exhibit very low current in the potential range from −0.45 to +0.6 V vs. Ag|AgCl|0.1 M KCl that reflects low capacitance related with low conductivity of titania and/or low surface area since we modify here the nanotubes of length below 1.6 µm. This current originates from charging/discharging of electric double layer. However, above this potential regime, oxygen evolution reaction (OER) was identified (marked as d, in Figure 9), while, below −0.45 V, the redox reactions are observed. Pairs of current peaks at −0.66 V(e)/−0.66 V(c), −0.85 V (f)/−0.85 V (b), and −0.99 V (g)/−0.96 V (a), versus Ag|AgCl|0.1 M KCl, could be ascribed to the redox reactions of $3Fe_2O_3 + 2e^- \leftrightarrow 2Fe_3O_4 + O^{2-}$, $TiO_2 + 2H_2O + e^- \leftrightarrow Ti(OH)_3 + OH^-$, and $Fe_3O_4 + 2H_3O^+ + 2e^- \leftrightarrow 3Fe(OH)_2$, respectively [67–71]. The redox pair (f/b) may also correspond to the oxidation and reduction reactions of the ions in the electrolyte with the TiO_2 film, and one possible reaction is $xNa + yTiO_2 + e^- \leftrightarrow Na_x(TiO_2)_y$ [72]. According to the Pourbaix diagram, any other forms of Ti in those pH conditions are not stable in the electrochemical system. The cathodic activity initiated at −1.1 V (h) and observed for Fe10L80 and Fe10L0 samples can be ascribed to the hydrogen evolution reaction (HER).

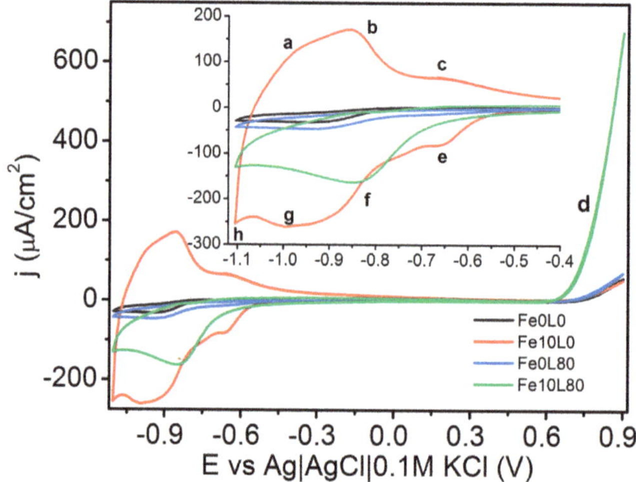

Figure 9. Cyclic voltammetry recorded for pristine and laser-modified; (80 mJ/cm²) TiO_2-NTs in 0.5 M NaOH electrolyte in darkness.

The comparison of the electrochemical performance of measured samples was carried out by the linear voltammetry technique performed in dark, under visible and full solar spectrum illumination, see Figure 10. In the potential range from −0.5 to + 0.7 V versus Ag|AgCl|0.1 M KCl the registered current of bare TiO_2-NTs is almost constant and highly sensitive for UV light. On the other hand, activity of Fe-decorated TiO_2-NTs samples is very low and almost light-independent until the polarization potential exceeds +0.1 V. Above this limit, the difference between registered current in dark and under light growths linearly up to beginning of OER, exceeding the photoactivity of bare TiO_2-NTs at +0.47, +0.56, and +0.75 V by Fe10L80, Fe10L120, and Fe10L160 samples, respectively. As shown in Figure 10, the current originating from OER in dark at +0.9 V versus Ag|AgCl|0.1 M KCl grows with the applied laser fluence from 50 µA/cm² for Fe10L0 to 744 µA/cm² for Fe10L80 and then decreases to 101 µA/cm² for Fe10L200. However, when the electrode is irradiated by visible light, the currents at +0.9 V grow up to 67, 995, and 128 µA/cm², for samples from Fe10 serie: un-annealed and treated with 10 and 80 mJ/cm², respectively. Only a few percent increments of registered currents were registered under UV-visible in reference to the results obtained under the visible radiation. Since the light wavelengths below 420 nm were blocked by the optical filter, the visible part of the spectrum has to be vastly responsible for the charge carriers' creation. The highest activity of Fe10L80 in dark compared with its relatively low light absorption ability leads to the conclusion that the light conversion efficiency is the main factor limiting the electrode performance, and the well-suited laser annealing can be employed to overcome that shortage. A presence of current peaks between −0.8 V and −1.1 V is seen only for Fe10L0 sample, supporting the CV conclusion about its prevailing derivation from iron oxidation reactions. It is worth noting that the presence of 10 nm Fe on the TiO$_2$-NTs leads to the photocurrent reduction under UV-visible light recorded for Fe10L0 and Fe10L40 samples (Figure 8), which shows that the presence of iron reduces the capability of electrodes to effectively transform light to charge carrier.

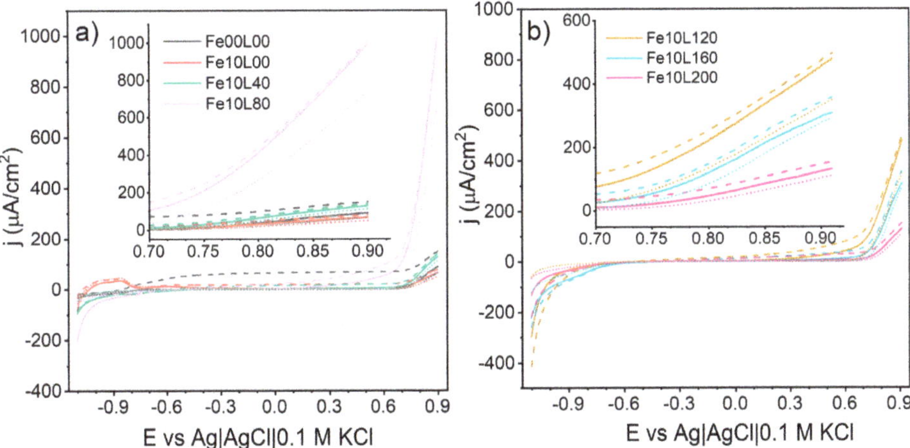

Figure 10. Records of linear voltammetry carried out in darkness (dotted), under simulated solar (dashed), and visible (solid) light for (**a**) bare titania and with Fe film further treated with 40 and 80 mJ/cm^2 and (**b**) 120,160, and 200 mJ/cm^2.

The stability of the laser-annealed Fe-decorated TiO$_2$-NTs was verified by the chronoamperometry during exposition to the chopped UV-visible and visible light registered at +0.4 V versus Ag|AgCl|0.1 M KCl. As shown in Figure 11a, the highest photocurrent density under UV-visible light was recorded for the bare sample (69 μA/cm^2), significantly dropped for Fe10L0 and Fe10L40 samples (18 μA/cm^2), and reached a local maximum for Fe10L80 (55 μA/cm^2) sample. The fluence of laser annealing above 80 mJ/cm^2 decreased the recorded currents. On the other hand, when the UV light was cut-off (Figure 11b), the highest photocurrent was registered for the Fe10L80 sample (22 μA/cm^2), while *j* of bare TiO$_2$-NTs reduces to 1.7 μA/cm^2. As shown in Figure 11c, the laser annealing of the bare TiO$_2$-NTs with a fluence of 40 mJ/cm^2 reduces registered photocurrent under UV-visible light, while the usage of the laser of 80 and 120 mJ/cm^2 shows local maximum of photoacitivy, similar to Fe-decorated samples. However, examination under exclusively visible light clearly shows that the Fe-decoration before laser processing is crucial to shift electrode photoactivity to the visible light.

The shape of the transient photocurrent curves under UV-vis illumination resembles the ideal square response only for three samples, namely Fe0L0, Fe10L80, and Fe10L200 (Figure 11a). In the case of Fe10L120 and Fe10L160 samples, upon illumination, the photocurrent rises instantly, but, when light is off, the photocurrent for a moment reverse the flow direction and stabilize at a few μA/cm^2. On the other hand, Fe10L0 and Fe10L40 samples, instead of current spikes, exhibit slowly asymptotically growing/decaying signal depending on the illumination variation. Relaxation current denotes a high density of surface states mostly attributed to the surface crystalline defects and oxygen vacancies acting as recombination centers for photo-generated carriers [73,74]. Thus, variation of electrodes behavior under solar irradiation highlights the differences in the electronic structure, especially between Fe10L40 and Fe10L80 samples.

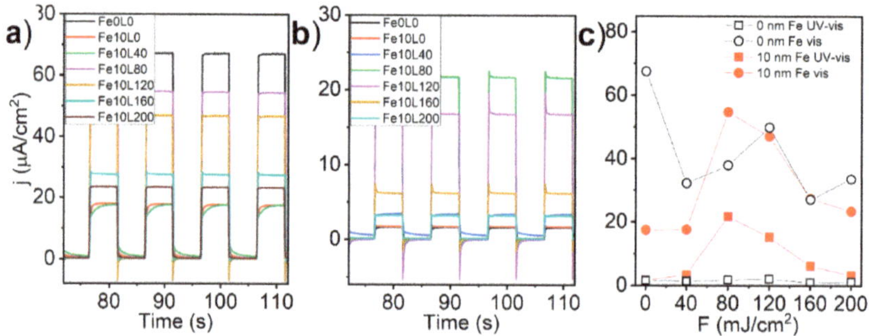

Figure 11. Transient photocurrents recorded for the bare and laser-annealed Fe-decorated TiO$_2$-NTs in 0.5 M NaOH electrolyte under the light of full solar (**a**) and limited to visible (**b**) spectrum. Light-on current plateau of chronoamperometry records in UV-vis (circle) and visible light (square) are shown in graph (**c**).

In order to investigate in details, the influence of the laser annealing on the charge carrier density and band position of iron decorated TiO$_2$-NTs, Mott-Schottky analysis was performed. Taking into account Equation (1), the N_d and E_{FB} were calculated from the linear part of the Mott-Shottky plot. As shown in Figure 12, thin layer of iron causes only a minor variation in TiO$_2$-NTs electrical properties, whenever the laser annealing cathodically shifts E_{FB} from −694 mV for Fe10L0 to −822 mV versus Ag|AgCl|0.1 M KCl for Fe10L200. On the other hand, N_d increases over 20 times from 1.71×10^{19} cm^{-3} for Fe10F0 to 36.4×10^{19} cm^{-3} for Fe10L120 but decreases for higher laser fluence. It is worthy of note that the most prominent change of N_d value takes place between Fe10L40 and Fe10L80 samples, showing the presence of processing threshold. That result corresponds with the XRD and Raman spectra variation between those materials, showing that the Fe10L80 sample exhibits the strongest structure distortion accompanied by oxygen vacancies formation. For a comparison, Haryński et al. [39] observed donor density increment from 2.2×10^{19} to 14.1×10^{20} cm^{-3} caused by the laser annealing using a fluence of 40 mJ/cm^2, while Xu et al. [16] observed from 3.9×10^{19} to 2.66×10^{20} cm^{-3} using 300 mJ/cm^2 but with KrF laser. In both cases, the authors modified TiO$_2$-NTs without any former deposition of other metal/metal oxide film. Thus, it is clear that the laser interaction with titania is mostly responsible for the charge carriers' concentration within the re-solidified layer.

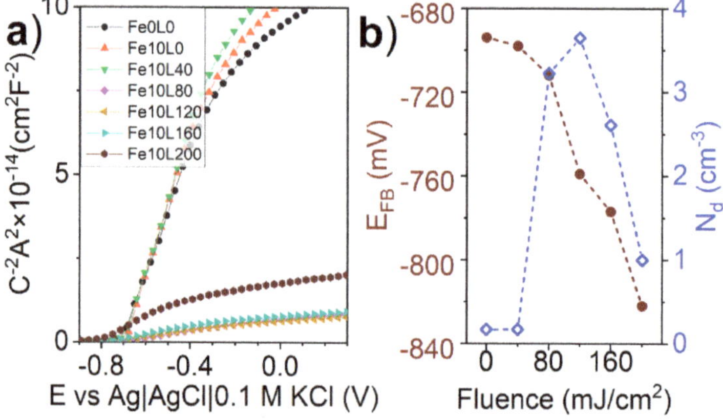

Figure 12. Mott-Schottky plots for bare and laser-annealed Fe-decorated TiO$_2$-NTs (**a**), and E_{FB} and N_d as a function of laser fluence (**b**).

It should be highlighted that the highest N_d was determined for the samples exhibiting most prominent transient current spikes (Figure 11b), both cathodic and anodic, under exposition to visible light. Since the capacitance of the double layer increases with the charge carriers density in the electrode, the presence of current spikes can be explained in terms of the Randles equivalent circuit of electrode-electrolyte interphase [75]. Turning on illumination causes the rapid growth of the charge carriers' density near the double layer followed by the increment of the junction capacitance (Equation (1)). Since the C_{SC} increases almost instantly, the additional charging current associated with accumulation of holes in the electrode space charge layer is observed as an anodic spike. However, when the light is turned off, the C_{SC} decreases and causes discharging current flowing in the opposite direction. The current originated from recombination of bulk electrons with holes accumulated in the space charge layer could inflict the potential even higher than the electrode bias potential; thus, the equipment records the unloading current as a reduction spike. It is worthy of note, that the ratio of discharging to the charging spikes is proportional to the hole transfer barrier across the electrode/electrolyte junction [76]. Therefore, the recorded transient photocurrent, especially for Fe10L120 and Fe10L80 (Figure 11), implies the reduction of the hole transfer effectiveness with increasing laser fluence.

Although the Fe10L120 sample exhibits the highest N_d, the Fe10L80 was measured to have the best electrochemical performance under the visible light. To explain that difference, the impact of the flat-band potential has to be taken into account. TiO_2-NTs are n-type semiconductor in which the Fermi level is higher than those of the water oxidation reaction in 0.5 NaOH electrolyte. Therefore, after immersion of the TiO_2-NTs in the electrolyte, it is starting to reach equilibrium between the Fermi levels giving a contribution to the band bending [77]. In general, the higher band bending the stronger transfer of holes towards the electrolyte, hampering their recombination with electrons. As a consequence, the negative shift of E_{FB} induced by the laser annealing reduces the band-bending and consequently reduces the oxidation under anodic but enhances reduction processes under cathodic polarization [78].

Another study of laser processing of TiO_2-NTs has shown that, if the laser annealing is carried out in vacuum [39], it leads to the positive shift of the E_{FB}, but it leads to the negative if the process was performed in water. On the other hand, it is known that the presence of hematite on the surface of TiO_2-NTs [79], as well as their doping by Fe, leads to the negative shift of the E_{FB} [80,81]. Taking into account the variation of E_{FB} as a laser fluence function (Figure 12), it should be concluded that the iron incorporation into the TiO_2 structure is responsible for the E_{FB} negative shift.

4. Conclusions

The work was focused onto the introduction of significant changes in morphology and structure of metal decorated titania via the well optimized laser treatment. Regarding the surface appearance, it has been shown that 1500 nm long TiO_2-NTs under the laser irradiation undergo partial melting and re-solidification leading to the formation of a continuous layer hiding the ordered tubes. The combination of anatase TiO_2, rutile TiO_2, and a hematite Fe_2O_3 has been revealed. However, anatase and rutile structures were found to be distorted due to the laser interaction. Additionally, a drastic decrement of UV light absorption leading to the disappearance of the optical bandgap of samples treated with the laser of fluence equal to or higher than 40 mJ/cm^2 was observed.

The presence of the processing threshold between samples Fe10L40 and Fe10L80 was confirmed by the rapid change in recorded transient photocurrent and the concentration of donor density, while, in the case of flat-band potential value, the cathodic shift was registered. It was proven that the charge carriers' density was strongly enhanced by the laser processing of the TiO_2-NTs but only up to laser fluence of 120 mJ/cm^2. Finally, the electrochemical performance which is determined here by the charge carriers' density, the position of Fermi level and light harvesting was found to be the highest for Fe10L80 sample. For that material the photocurrents recorded under visible light were over 20 times higher than for the bare TiO_2-NTs. The present study has shown that the photoactivity of

TiO$_2$-NTs electrode can be strongly enhanced by the Fe decoration and laser annealing under optimized conditions. This successful modification begins a new approach in laser assisted synthesis of decorated by transition metal TiO$_2$-NTs without the need to use any metal precursor in liquid form minimizing negative environmental impact.

Taking into account obtained results, it can be concluded that prepared samples can be used in photo-driven processes, e.g., in pollutant degradation or electrochemical water splitting enhanced by light. Moreover, Fe-TiO$_2$ nanotubes are also expected to be a promising alternative for cathodic protection of metals both under dark conditions and in visible light irradiation. It should be also mentioned that Fe species not only enhance the activity of titania nanotubes but also their presence can lead to good magnetic response under applied magnetic field. Therefore, such material can find application in biomedicine, e.g., when magnetically guidable systems are required. The range of possible application is very broad, and the fabrication route proposed here can definitely help with material production not only in laboratory but also in commercial scale.

Author Contributions: Conceptualization, K.S. and K.G.; methodology, K.S. and K.G.; validation, P.K., K.S. and K.G.; formal analysis, P.K., J.K. and J.W.; investigation, P.K., J.K. and J.W.; resources, K.S.; writing—original draft preparation, P.K.; writing—review and editing, K.S. and K.G.; visualization, P.K.; supervision, K.S. and K.G.; project administration, K.S.; funding acquisition, K.S. All authors have read and agreed to the published version of the manuscript.

Funding: This research was funded by The National Science Centre (Poland), grant number 2017/26/E/ST5/00416.

Conflicts of Interest: The authors declare no conflict of interest. The funders had no role in the design of the study; in the collection, analyses, or interpretation of data; in the writing of the manuscript, or in the decision to publish the results.

References

1. World Energy Outlook 2019—Analysis. Available online: https://www.iea.org/reports/world-energy-outlook-2019 (accessed on 4 May 2020).
2. Edwards, P.P.; Kuznetsov, V.L.; David, W.I.F. Hydrogen energy. *Philos. Trans. R. Soc. A* **2007**, *365*, 1043–1056. [CrossRef] [PubMed]
3. Chen, S.; Takata, T.; Domen, K. Particulate photocatalysts for overall water splitting. *Nat. Rev. Mater.* **2017**, *2*, 17050. [CrossRef]
4. Cook, T.R.; Dogutan, D.K.; Reece, S.Y.; Surendranath, Y.; Teets, T.S.; Nocera, D.G. Solar Energy Supply and Storage for the Legacy and Nonlegacy Worlds. *Chem. Rev.* **2010**, *110*, 6474–6502. [CrossRef] [PubMed]
5. Chu, S.; Li, W.; Yan, Y.; Hamann, T.; Shih, I.; Wang, D.; Mi, Z. Roadmap on solar water splitting: Current status and future prospects. *Nano Futures* **2017**, *1*, 022001. [CrossRef]
6. Chen, X.; Shen, S.; Guo, L.; Mao, S.S. Semiconductor-based Photocatalytic Hydrogen Generation. *Chem. Rev.* **2010**, *110*, 6503–6570. [CrossRef] [PubMed]
7. Van Benthem, K.; Elsässer, C.; French, R.H. Bulk electronic structure of SrTiO$_3$: Experiment and theory. *J. Appl. Phys.* **2001**, *90*, 6156–6164. [CrossRef]
8. Xu, Y.; Schoonen, M.A.A. The absolute energy positions of conduction and valence bands of selected semiconducting minerals. *Am. Mineral.* **2000**, *85*, 543–556. [CrossRef]
9. Syama, S.; Reshma, S.C.; Sreekannth, P.J.; Varma, H.K.; Mohanan, P.V. Effect of zinc oxide nanoparticles on cellular oxidative stress and antioxidant defence mechanisms in mouse liver. *Environ. Toxicol. Chem.* **2013**, *95*, 495–503. [CrossRef]
10. Makumire, S.; Chakravadhanula, V.S.K.; Kollisch, G.; Redel, E.; Shonhai, A. Immunomodulatory activity of zinc peroxide (ZnO$_2$) and titanium dioxide (TiO$_2$) nanoparticles and their effects on DNA and protein integrity. *Toxicol. Lett.* **2014**, *227*, 56–64. [CrossRef]
11. Fujishima, A.; Honda, K. Electrochemical Photolysis of Water at a Semiconductor Electrode. *Nature* **1972**, *238*, 37–38. [CrossRef]
12. O'Regan, B.; Grätzel, M. A low-cost, high-efficiency solar cell based on dye-sensitized colloidal TiO$_2$ films. *Nature* **1991**, *353*, 737–740. [CrossRef]
13. Walter, M.G.; Warren, E.L.; McKone, J.R.; Boettcher, S.W.; Mi, Q.; Santori, E.A.; Lewis, N.S. Solar Water Splitting Cells. *Chem. Rev.* **2010**, *110*, 6446–6473. [CrossRef] [PubMed]

14. Zwilling, V.; Aucouturier, M.; Darque-Ceretti, E. Anodic oxidation of titanium and TA6V alloy in chromic media. An electrochemical approach. *Electrochim. Acta* **1999**, *45*, 921–929. [CrossRef]

15. Fu, Y.; Mo, A. A Review on the Electrochemically Self-organized Titania Nanotube Arrays: Synthesis, Modifications, and Biomedical Applications. *Nanoscale Res. Lett.* **2018**, *13*, 187. [CrossRef] [PubMed]

16. Xu, Y.; Melia, M.A.; Tsui, L.; Fitz-Gerald, J.M.; Zangari, G. Laser-Induced Surface Modification at Anatase TiO_2 Nanotube Array Photoanodes for Photoelectrochemical Water Oxidation. *J. Phys. Chem. C* **2017**, *121*, 17121–17128. [CrossRef]

17. Siuzdak, K.; Szkoda, M.; Sawczak, M.; Karczewski, J.; Ryl, J.; Cenian, A. Ordered titania nanotubes layer selectively annealed by laser beam for high contrast electrochromic switching. *Thin Solid Films* **2018**, *659*, 48–56. [CrossRef]

18. Liang, Z.; Hou, H.; Fang, Z.; Gao, F.; Wang, L.; Chen, D.; Yang, W. Hydrogenated TiO_2 Nanorod Arrays Decorated with Carbon Quantum Dots toward Efficient Photoelectrochemical Water Splitting. *ACS Appl. Mater. Interfaces* **2019**, *11*, 19167–19175. [CrossRef]

19. Meng, A.; Zhang, J.; Xu, D.; Cheng, B.; Yu, J. Enhanced photocatalytic H_2 -production activity of anatase TiO_2 nanosheet by selectively depositing dual-cocatalysts on {101} and {001} facets. *Appl. Catal. B Environ.* **2016**, *198*, 286–294. [CrossRef]

20. Chiarello, G.L.; Zuliani, A.; Ceresoli, D.; Martinazzo, R.; Selli, E. Exploiting the Photonic Crystal Properties of TiO_2 Nanotube Arrays To Enhance Photocatalytic Hydrogen Production. *ACS Catal.* **2016**, *6*, 1345–1353. [CrossRef]

21. Zu, G.; Li, H.; Liu, S.; Li, D.; Wang, J.; Zhao, J. Highly efficient mass determination of TiO_2 nanotube arrays and its application in lithium-ion batteries. *Sustain. Mater. Technol.* **2018**, *18*, e00079. [CrossRef]

22. Ozkan, S.; Nguyen, N.T.; Mazare, A.; Schmuki, P. Optimized Spacing between TiO_2 Nanotubes for Enhanced Light Harvesting and Charge Transfer. *ChemElectroChem* **2018**, *5*, 3183–3190. [CrossRef]

23. Wang, M.; Sun, L.; Cai, J.; Huang, P.; Su, Y.; Lin, C. A facile hydrothermal deposition of $ZnFe_2O_4$ nanoparticles on TiO_2 nanotube arrays for enhanced visible light photocatalytic activity. *J. Mater. Chem. A* **2013**, *1*, 12082. [CrossRef]

24. Kuang, S.; Yang, L.; Luo, S.; Cai, Q. Fabrication, characterization and photoelectrochemical properties of Fe_2O_3 modified TiO_2 nanotube arrays. *Appl. Surf. Sci.* **2009**, *255*, 7385–7388. [CrossRef]

25. Litter, M.I.; Navío, J.A. Comparison of the photocatalytic efficiency of TiO_2, iron oxides and mixed Ti(IV) Fe(III) oxides: Photodegradation of oligocarboxylic acids. *J. Photochem. Photobiol. A Chem.* **1994**, *84*, 183–193. [CrossRef]

26. Molenda, Z.; Grochowska, K.; Karczewski, J.; Ryl, J.; Darowicki, K.; Rysz, J.; Cenian, A.; Siuzdak, K. The influence of the Cu_2O deposition method on the structure, morphology and photoresponse of the ordered TiO_2NTs/Cu_2O heterojunction. *Mater. Res. Express* **2020**, *6*, 1250b6. [CrossRef]

27. Wu, Z.; Wang, Y.; Sun, L.; Mao, Y.; Wang, M.; Lin, C. An ultrasound-assisted deposition of NiO nanoparticles on TiO_2 nanotube arrays for enhanced photocatalytic activity. *J. Mater. Chem. A* **2014**, *2*, 8223. [CrossRef]

28. Roy, P.; Kim, D.; Paramasivam, I.; Schmuki, P. Improved efficiency of TiO_2 nanotubes in dye sensitized solar cells by decoration with TiO_2 nanoparticles. *Electrochem. Commun.* **2009**, *11*, 1001–1004. [CrossRef]

29. Chen, S.; Paulose, M.; Ruan, C.; Mor, G.K.; Varghese, O.K.; Kouzoudis, D.; Grimes, C.A. Electrochemically synthesized CdS nanoparticle-modified TiO_2 nanotube-array photoelectrodes: Preparation, characterization, and application to photoelectrochemical cells. *J. Photochem. Photobiol. A Chem.* **2006**, *177*, 177–184. [CrossRef]

30. Assaud, L.; Brazeau, N.; Barr, M.K.S.; Hanbücken, M.; Ntais, S.; Baranova, E.A.; Santinacci, L. Atomic Layer Deposition of Pd Nanoparticles on TiO_2 Nanotubes for Ethanol Electrooxidation: Synthesis and Electrochemical Properties. *ACS Appl. Mater. Interfaces* **2015**, *7*, 24533–24542. [CrossRef]

31. Zhao, L.; Wang, H.; Huo, K.; Cui, L.; Zhang, W.; Ni, H.; Zhang, Y.; Wu, Z.; Chu, P.K. Antibacterial nano-structured titania coating incorporated with silver nanoparticles. *Biomaterials* **2011**, *32*, 5706–5716. [CrossRef]

32. Han, H.; Riboni, F.; Karlicky, F.; Kment, S.; Goswami, A.; Sudhagar, P.; Yoo, J.; Wang, L.; Tomanec, O.; Petr, M.; et al. α-Fe_2O_3/TiO_2 3D hierarchical nanostructures for enhanced photoelectrochemical water splitting. *Nanoscale* **2017**, *9*, 134–142. [CrossRef]

33. Kurien, U.; Hu, Z.; Lee, H.; Dastoor, A.P.; Ariya, P.A. Radiation enhanced uptake of HgO(g) on iron (oxyhydr)oxide nanoparticles. *RSC Adv.* **2017**, *7*, 45010–45021. [CrossRef]

34. Kment, S.; Riboni, F.; Pausova, S.; Wang, L.; Wang, L.; Han, H.; Hubicka, Z.; Krysa, J.; Schmuki, P.; Zboril, R. Photoanodes based on TiO_2 and α-Fe_2O_3 for solar water splitting–superior role of 1D nanoarchitectures and of combined heterostructures. *Chem. Soc. Rev.* **2017**, *46*, 3716–3769. [CrossRef]

35. Lin, Y.-G.; Hsu, Y.-K.; Lin, Y.-C.; Chen, Y.-C. Electrodeposited Fe_2TiO_5 nanostructures for photoelectrochemical oxidation of water. *Electrochim. Acta* **2016**, *213*, 898–903. [CrossRef]

36. Bassi, P.S.; Antony, R.P.; Boix, P.P.; Fang, Y.; Barber, J.; Wong, L.H. Crystalline Fe_2O_3/Fe_2TiO_5 heterojunction nanorods with efficient charge separation and hole injection as photoanode for solar water oxidation. *Nano Energy* **2016**, *22*, 310–318. [CrossRef]

37. Hsu, M.-Y.; Van Thang, N.; Wang, C.; Leu, J. Structural and morphological transformations of TiO_2 nanotube arrays induced by excimer laser treatment. *Thin Solid Films* **2012**, *520*, 3593–3599. [CrossRef]

38. Kusinski, J.; Kac, S.; Kopia, A.; Radziszewska, A.; Rozmus-Górnikowska, M.; Major, B.; Major, L.; Marczak, J.; Lisiecki, A. Laser modification of the materials surface layer—A review paper. *B Pol. Acad. Sci-Tech.* **2012**, *60*, 711–728. [CrossRef]

39. Haryński, Ł.; Grochowska, K.; Karczewski, J.; Ryl, J.; Siuzdak, K. Scalable Route toward Superior Photoresponse of UV-Laser-Treated TiO_2 Nanotubes. *ACS Appl. Mater. Interfaces* **2020**, *12*, 3225–3235. [CrossRef]

40. Dholam, R.; Patel, N.; Adami, M.; Miotello, A. Hydrogen production by photocatalytic water-splitting using Cr- or Fe-doped TiO_2 composite thin films photocatalyst. *Int. J. Hydrogen Energy* **2009**, *34*, 5337–5346. [CrossRef]

41. Xu, C.; Zhang, Y.; Chen, J.; Lin, J.; Zhang, X.; Wang, Z.; Zhou, J. Enhanced mechanism of the photo-thermochemical cycle based on effective Fe-doping TiO_2 films and DFT calculations. *Appl. Catal. B Environ.* **2017**, *204*, 324–334. [CrossRef]

42. Pradubkorn, P.; Maensiri, S.; Swatsitang, E.; Laokul, P. Preparation and characterization of hollow TiO_2 nanospheres: The effect of Fe^{3+} doping on their microstructure and electronic structure. *Curr. Appl. Phys.* **2020**, *20*, 178–185. [CrossRef]

43. Ali, G.; Chen, C.; Yoo, S.H.; Kum, J.M.; Cho, S.O. Fabrication of complete titania nanoporous structures via electrochemical anodization of Ti. *Nanoscale Res. Lett.* **2011**, *6*, 332. [CrossRef]

44. Swift, P. Adventitious carbon—The panacea for energy referencing? *Surf. Interface Anal.* **1982**, *4*, 47–51. [CrossRef]

45. Gelderman, K.; Lee, L.; Donne, S.W. Flat-Band Potential of a Semiconductor: Using the Mott–Schottky Equation. *J. Chem. Educ.* **2007**, *84*, 685. [CrossRef]

46. Park, B.H.; Li, L.S.; Gibbons, B.J.; Huang, J.Y.; Jia, Q.X. Photovoltaic response and dielectric properties of epitaxial anatase-TiO_2 films grown on conductive $La_{0.5}Sr_{0.5}CoO_3$ electrodes. *Appl. Phys. Lett.* **2001**, *79*, 2797–2799. [CrossRef]

47. Wawrzyniak, J.; Karczewski, J.; Kupracz, P.; Grochowska, K.; Załeski, K.; Pshyk, O.; Coy, E.; Bartmański, M.; Szkodo, M.; Siuzdak, K. Laser-assisted modification of titanium dioxide nanotubes in a tilted mode as surface modification and patterning strategy. *Appl. Surf. Sci.* **2020**, *508*, 145143. [CrossRef]

48. Werner, W.S.M.; Glantschnig, K.; Ambrosch-Draxl, C. Optical Constants and Inelastic Electron-Scattering Data for 17 Elemental Metals. *J. Phys. Chem. Ref. Data* **2009**, *38*, 1013–1092. [CrossRef]

49. Querry, M.R. Optical Constants, Contractor Report CRDC-CR-85034. 1985. Available online: https://apps.dtic.mil/dtic/tr/fulltext/u2/a158623.pdf (accessed on 9 September 2020).

50. Siefke, T.; Kroker, S.; Pfeiffer, K.; Puffky, O.; Dietrich, K.; Franta, D.; Ohlídal, I.; Szeghalmi, A.; Kley, E.-B.; Tünnermann, A. Materials Pushing the Application Limits of Wire Grid Polarizers further into the Deep Ultraviolet Spectral Range. *Adv. Opt. Mater.* **2016**, *4*, 1780–1786. [CrossRef]

51. Li Bassi, A.; Cattaneo, D.; Russo, V.; Bottani, C.E.; Barborini, E.; Mazza, T.; Piseri, P.; Milani, P.; Ernst, F.O.; Wegner, K.; et al. Raman spectroscopy characterization of titania nanoparticles produced by flame pyrolysis: The influence of size and stoichiometry. *J. Appl. Phys.* **2005**, *98*, 074305. [CrossRef]

52. Ohsaka, T.; Izumi, F.; Fujiki, Y. Raman spectrum of anatase, TiO_2. *J. Raman Spectrosc.* **1978**, *7*, 321–324. [CrossRef]

53. Narayanan, P.S. Raman spectrum of rutile (TiO_2). *Proc. Indian Acad. Sci. (Math. Sci.)* **1950**, *32*, 279. [CrossRef]

54. De Faria, D.L.A.; Silva, S.V.; de Oliveira, M.T. Raman microspectroscopy of some iron oxides and oxyhydroxides. *J. Raman Spectrosc.* **1997**, *28*, 873–878. [CrossRef]

55. McCarty, K.F. Inelastic light scattering in α-Fe_2O_3: Phonon vs magnon scattering. *Solid State Commun.* **1988**, *68*, 799–802. [CrossRef]

56. Shebanova, O.N.; Lazor, P. Raman spectroscopic study of magnetite ($FeFe_2O_4$): A new assignment for the vibrational spectrum. *J. Solid State Chem.* **2003**, *174*, 424–430. [CrossRef]

57. Wang, B.; Shen, S.; Mao, S.S. Black TiO_2 for solar hydrogen conversion. *J. Mater.* **2017**, *3*, 96–111. [CrossRef]

58. Wang, X.H.; Li, J.-G.; Kamiyama, H.; Katada, M.; Ohashi, N.; Moriyoshi, Y.; Ishigaki, T. Pyrogenic Iron(III)-Doped TiO_2 Nanopowders Synthesized in RF Thermal Plasma: Phase Formation, Defect Structure, Band Gap, and Magnetic Properties. *J. Am. Chem. Soc.* **2005**, *127*, 10982–10990. [CrossRef]

59. Lipińska, W.; Siuzdak, K.; Ryl, J.; Barski, P.; Śliwiński, G.; Grochowska, K. The optimization of enzyme immobilization at Au-Ti nanotextured platform and its impact onto the response towards glucose in neutral media. *Mater. Res. Express* **2019**, *6*, 1150e3. [CrossRef]

60. Liu, Y.-T.; Yuan, Q.-B.; Duan, D.-H.; Zhang, Z.-L.; Hao, X.-G.; Wei, G.-Q.; Liu, S.-B. Electrochemical activity and stability of core–shell Fe_2O_3/Pt nanoparticles for methanol oxidation. *J. Power Sources* **2013**, *243*, 622–629. [CrossRef]

61. Biesinger, M.C.; Payne, B.P.; Grosvenor, A.P.; Lau, L.W.M.; Gerson, A.R.; Smart, R.S.C. Resolving surface chemical states in XPS analysis of first row transition metals, oxides and hydroxides: Cr, Mn, Fe, Co and Ni. *Appl. Surf. Sci.* **2011**, *257*, 2717–2730. [CrossRef]

62. Gilbert, B.; Frandsen, C.; Maxey, E.R.; Sherman, D.M. Band-gap measurements of bulk and nanoscale hematite by soft x-ray spectroscopy. *Phys. Rev. B* **2009**, *79*, 035108. [CrossRef]

63. Lany, S. Band-structure calculations for the 3d transition metal oxides in G W. *Phys. Rev. B* **2013**, *87*, 085112. [CrossRef]

64. Persson, K. *Materials Data on TiFeO3 (SG:148) by Materials Project*; Office of Scientific and Technical Information (OSTI): Oak Ridge, TN, USA, 2014. [CrossRef]

65. Persson, K. *Materials Data on Ti(FeO2)2 (SG:15) by Materials Project*; Office of Scientific and Technical Information (OSTI): Oak Ridge, TN, USA, 2014. [CrossRef]

66. Persson, K. *Materials Data on TiFe2O5 (SG:15) by Materials Project*; Office of Scientific and Technical Information (OSTI): Oak Ridge, TN, USA, 2020. [CrossRef]

67. Macak, J.M.; Tsuchiya, H.; Ghicov, A.; Yasuda, K.; Hahn, R.; Bauer, S.; Schmuki, P. TiO_2 nanotubes: Self-organized electrochemical formation, properties and applications. *Curr. Opin. Solid State Mater. Sci.* **2007**, *11*, 3–18. [CrossRef]

68. Schmuki, P. Passivity of Iron in Alkaline Solutions Studied by In Situ XANES and a Laser Reflection Technique. *J. Electrochem. Soc.* **1999**, *146*, 2097. [CrossRef]

69. Neugebauer, H.; Moser, A.; Strecha, P.; Neckel, A. In Situ FTIR Spectroscopy of Iron Electrodes in Alkaline Solutions. *J. Electrochem. Soc.* **1990**, *137*, 4. [CrossRef]

70. Kavan, L.; Grätzel, M.; Rathouský, J.; Zukalb, A. Nanocrystalline TiO_2 (Anatase) Electrodes: Surface Morphology, Adsorption, and Electrochemical Properties. *J. Electrochem. Soc.* **1996**, *143*, 394. [CrossRef]

71. Bülter, H.; Denuault, G.; Mátéfi-Tempfli, S.; Mátéfi-Tempfli, M.; Dosche, C.; Wittstock, G. Electrochemical analysis of nanostructured iron oxides using cyclic voltammetry and scanning electrochemical microscopy. *Electrochim. Acta* **2016**, *222*, 1326–1334. [CrossRef]

72. Sun, X.; Xie, M.; Travis, J.J.; Wang, G.; Sun, H.; Lian, J.; George, S.M. Pseudocapacitance of amorphous TiO_2 thin films anchored to graphene and carbon nanotubes using atomic layer deposition. *J. Phys. Chem. C* **2013**, *117*, 44–22497. [CrossRef]

73. Krysa, J.; Zlamal, M.; Kment, S.; Brunclikova, M.; Hubicka, Z. TiO_2 and Fe_2O_3 Films for Photoelectrochemical Water Splitting. *Molecules* **2015**, *20*, 1046–1058. [CrossRef]

74. Sivula, K. Metal Oxide Photoelectrodes for Solar Fuel Production, Surface Traps, and Catalysis. *J. Phys. Chem. Lett.* **2013**, *4*, 1624–1633. [CrossRef]

75. Randles, J.E.B. Kinetics of rapid electrode reactions. *Discuss. Faraday Soc.* **1947**, *1*, 11. [CrossRef]

76. Tamirat, A.G.; Dubale, A.A.; Su, W.-N.; Chen, H.-M.; Hwang, B.-J. Sequentially surface modified hematite enables lower applied bias photoelectrochemical water splitting. *Phys. Chem. Chem. Phys.* **2017**, *19*, 20881–20890. [CrossRef] [PubMed]

77. Cheng, X.; Kong, D.-S.; Wang, Z.; Feng, Y.-Y.; Li, W.-J. Inhibiting effect of carbonate on the photoinduced flatband potential shifts during water photooxidation at TiO_2/solution interface. *J. Solid State Electrochem.* **2017**, *21*, 1467–1475. [CrossRef]

78. Jackman, M.J.; Thomas, A.G.; Muryn, C. Photoelectron Spectroscopy Study of Stoichiometric and Reduced Anatase TiO_2 (101) Surfaces: The Effect of Subsurface Defects on Water Adsorption at Near-Ambient Pressures. *J. Phys. Chem. C* **2015**, *119*, 13682–13690. [CrossRef]

79. Sołtys-Mróz, M.; Syrek, K.; Pierzchała, J.; Wiercigroch, E.; Malek, K.; Sulka, G.D. Band gap engineering of nanotubular Fe_2O_3-TiO_2 photoanodes by wet impregnation. *Appl. Surf. Sci.* **2020**, *517*, 146195. [CrossRef]

80. Ganesh, I.; Kumar, P.; Gupta, A.; Sekhar, P.; Radha, K.; Padmanabham, G.; Sundararajan, G. Preparation and characterization of Fe-doped TiO_2 powders for solar light response and photocatalytic applications. *PAC* **2012**, *6*, 21–36. [CrossRef]

81. Lee, T.; Ryu, H.; Lee, W.-J. Photoelectrochemical properties of iron (III)-doped TiO_2 nanorods. *Ceram. Int.* **2015**, *41*, 7582–7589. [CrossRef]

 *materials*

Article

Ag-Decorated Si Microspheres Produced by Laser Ablation in Liquid: All-in-One Temperature-Feedback SERS-Based Platform for Nanosensing

Stanislav Gurbatov [1,2], Vladislav Puzikov [1], Evgeny Modin [3], Alexander Shevlyagin [1], Andrey Gerasimenko [4], Eugeny Mitsai [1], Sergei A. Kulinich [5] and Aleksandr Kuchmizhak [1,2,*]

[1] Institute of Automation and Control Processes, Far Eastern Branch, Russian Academy of Science, 5 Radio Str., 690041 Vladivostok, Russia
[2] Far Eastern Federal University, 690041 Vladivostok, Russia
[3] CIC NanoGUNE BRTA, 20018 Donostia-San Sebastian, Spain
[4] Institute of Chemistry, Far Eastern Branch, Russian Academy of Science, 690022 Vladivostok, Russia
[5] Research Institute of Science & Technology, Tokai University, Hiratsuka 259-1292, Kanagawa, Japan
* Correspondence: alex.iacp.dvo@mail.ru

Citation: Gurbatov, S.; Puzikov, V.; Modin, E.; Shevlyagin, A.; Gerasimenko, A.; Mitsai, E.; Kulinich, S.A.; Kuchmizhak, A. Ag-Decorated Si Microspheres Produced by Laser Ablation in Liquid: All-in-One Temperature-Feedback SERS-Based Platform for Nanosensing. *Materials* **2022**, *15*, 8091. https://doi.org/10.3390/ma15228091

Academic Editor: Alina A. Manshina

Received: 24 October 2022
Accepted: 12 November 2022
Published: 15 November 2022

Publisher's Note: MDPI stays neutral with regard to jurisdictional claims in published maps and institutional affiliations.

Abstract: Combination of dissimilar materials such as noble metals and common semiconductors within unified nanomaterials holds promise for optoelectronics, catalysis and optical sensing. Meanwhile, difficulty of obtaining such hybrid nanomaterials using common lithography-based techniques stimulates an active search for advanced, inexpensive, and straightforward fabrication methods. Here, we report one-pot one-step synthesis of Ag-decorated Si microspheres via nanosecond laser ablation of monocrystalline silicon in isopropanol containing $AgNO_3$. Laser ablation of bulk silicon creates the suspension of the Si microspheres that host further preferential growth of Ag nanoclusters on their surface upon thermal-induced decomposition of $AgNO_3$ species by subsequently incident laser pulses. The amount of the $AgNO_3$ in the working solution controls the density, morphology, and arrangement of the Ag nanoclusters allowing them to achieve strong and uniform decoration of the Si microsphere surface. Such unique morphology makes Ag-decorated Si microspheres promising for molecular identification based on the surface-enhanced Raman scattering (SERS) effect. In particular, the designed single-particles sensing platform was shown to offer temperature-feedback modality as well as SERS signal enhancement up to 10^6, allowing reliable detection of the adsorbed molecules and tracing their plasmon-driven catalytic transformations. Considering the ability to control the decoration degree of Si microspheres by Ag nanoclusters via amount of the $AgNO_3$, the developed one-pot easy-to-implement PLAL synthesis holds promise for gram-scale production of high-quality hybrid nanomaterial for various nanophotonics and sensing applications.

Keywords: pulsed laser ablation in liquid; hybrid nanomaterials; silver; silicon; plasmonics; SERS

1. Introduction

Noble-metal nanoparticles supporting optically induced resonance oscillations of their electron density (surface plasmons), as well as submicron structures made of low-loss highly refractive-index semiconductors supporting geometric Mie modes, have been the mainstream during last decades of both fundamental and applied research in the fields of nanophotonics, plasmonics, optoelectronics, catalyst, and optical sensing [1,2]. Progressive popularity of such functional nanomaterials is associated with their unique optical properties. In particular, surface plasmons offer strong light localization and enhancement at nanoscale volumes, while Mie modes permit to control radiation directivity and confine light within the related all-dielectric nanostructure boosting its inherent linear and nonlinear optical responses. Merging plasmonic and all-dielectric concepts within the unified hybrid nanostructures can be considered as a new step towards design of the advanced materials with expanded functionality [3,4]. In particular, plasmon-mediated highly energetic (also referred to as "hot") electrons can be injected to semiconductor to enhance its light-emitting and light-absorbing properties, allowing it to realize advanced photodetectors and nanoscale coherent light emitters [5–7]. Moreover, combination of noble metals and semiconductors holds promise for next-generation catalyst [8] as well as realization of multi-functional optical sensing platforms benefiting from plasmon-mediated enhancement of molecular fingerprint signals via surface–enhanced photoluminescence and surface-enhanced Raman scattering (SERS) [9] and chemical inertness of all-dielectric nanostructures [10]. Moreover, monocrystalline semiconductors usually exhibit inherent characteristic Raman scattering that can be enhanced by Mie resonances [11] and can be used for local nanothermometry [12]. However, it should be noted that the optical-range resonant response for plasmonic and all-dielectric nanostructures is usually observed for plasmonic and all-dielectric nanostructures with a rather dissimilar size. This issue makes application of common multi-step lithography-based nanofabrication techniques for hybrid nanostructure formation challenging and money-consuming, stimulating the search for more straightforward, productive, and versatile fabrication approaches.

Pulsed laser ablation in liquids (PLAL) can provide an easy-to-realize and elegant way to tackle the problem of formation of hybrid metal-semiconductor nanostructures [13–15]. In particular, popular combinations of metals and semiconductors mixed within the functional nanomaterials via PLAL synthesis include Ag- (Au- or Pt-) decorated TiO_2 [16–20] or Au-decorated Si [21–26]. These hybrid nanomaterials proved their usefulness for optical and gas sensing, catalysis, and nanophotonics, justifying PLAP as advanced and versatile fabrication technology. More specifically, PLAL can be considered a "chemically green" procedure, that provides rather unique synthesis conditions such as high temperature and GPa pressure gradients. This permits the production of systems with unique morphology [27–34], structure, and composition, including rarely observed nano-alloys and nanomaterials containing meta-stable phases [20,35,36]. Simplicity of the methods, its scalability, and diversity of source materials for nanoparticles formation, as well as deepening insight into the physics of laser-matter interaction make PLAL synthesis attractive for multigram-scale production of inexpensive nanomaterials for various practical applications.

Enormous efforts were made so far to obtain nanomaterials made of pure noble metals [31,37–39] using PLAL, while only a few studies attempted to adopt this method for combining noble metals with common semiconductor—silicon—within unified hybrid nanostructrues. In particular, Au-Si hybrids with various morphologies (such as core-shell, core-satellites, nanoalloy, nanosponges, etc.) were reported by several research groups highlighting their unique optical and nonlinear optical properties as well as high potential for nanoscale light emission, SERS-based biosensing, light-to-heat conversion, and phototherapy [21–26]. Saraeva, et al. demonstrated general applicability of PLAL-synthesis to produce Si microspheres decorated by Ag nanoclusters [40], yet structural features of the produced nanomaterial, its optical properties, and related applications were left behind.

In this study, high-quality nanomaterial in the form of Ag-decorated Si microspheres was produced upon single-step nanosecond (ns)-laser ablation of monocrystalline silicon in isopropanol containing optimized content of $AgNO_3$ precursor. At fixed laser irradiation time, salt content in the working solution was found to define the decoration degree with Ag nanoclusters that preferentially grow on the Si surface via thermal-induced decomposition of the precursor molecules. Morphology and inner structure of the as-prepared Ag-Si hybrids were comprehensively characterized by combining state-of-the-art 3D tomographic reconstruction with scanning electron microscopy (SEM) and focused ion-beam (FIB) milling, transmission electron microscopy (TEM) with energy-dispersive X-ray (EDX) chemical mapping modality as well as X-ray diffraction (XRD). Finally, our studies revealed applicability of the produced Ag-decorated microspheres as a versatile SERS platform offering temperature-feedback modality as well as single-particle signal enhancement that is well enough for fingerprint identification of the analyte molecules and tracing their plasmon-driven catalytic transformations.

2. Materials and Methods

2.1. PLAL Synthesis

Hybrid Ag-Si microspheres were produced via single-step laser ablation of monocrystalline Si wafer in isopropanol (purity grade > 99.9 wt.%, Sigma-Aldrich) containing silver nitrate ($AgNO_3$, purity 99.9%, Sigma-Aldrich, St. Louis, MO, USA). Non-doped (>2000 $\Omega\cdot$cm) Si wafer with (001) surface orientation was used in all laser ablation experiments. Prior to PLAL synthesis, Si target was subjected to a standard RCA cleaning procedure followed by a short dip in hydrofluoric acid and deionized water rinsing. In particular, Si wafer was fixed in a quartz cuvette containing a working solution made by mixing 9 mL of isopropanol and 1 mL of aqueous solution of $AgNO_3$. To control the decoration degree of the product, the amount of the $AgNO_3$ was varied from 5×10^{-5} to 10^{-2} M. Laser pulses (pulse duration of 7 ns, wavelength of 532 nm, and pulse repetition rate of 20 Hz) generated by Nd:YAG laser source (Ultra, Quantel, Lannion, France) at fixed pulse energy of $\approx$1.5 mJ were focused by a lens with a focal distance of 20 cm on the Si wafer surface. Such focusing parameters allow to ablate the circular-shape area ($\approx$0.13 mm^2) upon continuous single-spot irradiation. In all the experiments, laser exposure was carried out for 2 h upon continuous steering of the working solution.

2.2. Characterization

Morphology and inner structure of the Ag-Si hybrids were studied using SEM combined with a FIB milling modality (Helios 450s Nanolab, Thermofisher, Waltham, MA, USA). FIB milling of Pt-protected microspheres was carried out at 30 keV and a beam current of 25 pA to prepare their thin lamella for high-resolution TEM (HR-TEM) as well as to produce multiple slices for 3D tomographic reconstruction. The later procedure was performed with FEI AutoSlice & View G3 software (Thermofisher, USA), resulting in fabrication of about 110 slice images of the isolated nanostructure with an average slice thickness of 10 nm. The slice images were further processed with Avizo 8.1 software (Thermofisher, USA) to create the 3D model of the microsphere. Scanning TEM (STEM) imaging combined with an EDX spectroscopy (EDAX) for chemical mapping was carried out at an acceleration voltage of 300 kV (Titan 60–300, Thermo Fisher, USA).

To determine the phase composition and crystallinity of the Ag-Si microspheres supported by crystalline Si substrate, XRD method (Rigaku SmartLab, Tokyo, Japan) with Cu Kα radiation and parallel beam optics in $2\theta/\omega$ mode was used. The XRD peaks were identified using the ICSD database.

2.3. Applications

SERS experiments were carried out using commercial confocal laser system (Ntegra Spectra II, NT-MDT, Russia) equipped with CW laser sources (wavelengths λ_{pump} of 473 and 633 nm), focusing system with a dry microscope objective having numerical aperture (NA) of 0.7 (100×, M Plan Apo, Mitutoyo, Kawasaki, Japan), an avalanche photodetector for reflectivity mapping as well as an optical spectrometer with a monochromator (M522, Solar Laser Instruments, Belarus) and a thermo-electrically cooled CCD camera (i-Dus, Oxford Instruments, UK). An additional photodetector was used to control the laser intensity at the output of the focusing objective. The focal depth estimated as $\approx \lambda_{pump}/NA^2 \approx 0.96$ μm (at $\lambda_{pump} \approx 473$ nm) ensured uniform excitation of the microspheres on smooth substrates. Lateral size of the focal spot ($\approx 1.22\,\lambda_{pump}/NA$) was about 0.8 and 1.1 μm at λ_{pump} = 473 and 633 nm, respectively, also matching the Au-Si microsphere size. Prior to SERS measurements, Ag-Si microspheres were maintained in the isopropanol solutions containing Raman-active molecules, Rhodamine 6G (R6G), and para-aminothiophenol (PATP), at their molar concentrations of 10^{-6} M for 1 h. Then, the microspheres were cleaned with distilled water several times and drop-casted onto a smooth Ag mirror. Such mirrors were produced by covering monocrystalline Si wafer with a 500 nm thick silver film using magnetron sputtering. The concentration of the PLAL-synthesized nanomaterial in the distilled water was chosen to provide separated microspheres on the surface upon drying. All Raman and SERS measurements were performed on isolated Ag-Si microspheres.

Finite-difference time-domain (FDTD) simulations were undertaken with a commercial electromagnetic solver (Lumerical Solutions, Ansys, Canonsburg, PA, USA) to calculate the local structure of the electromagnetic fields in the vicinity of Ag-Si microspheres upon their excitation with various pump wavelengths ranging from 405 to 633 nm. The dielectric constants of silver and silicon at corresponding wavelength were taken from [41]. Simulations were carried out considering isolated Ag-Si microsphere with the precise geometrical model taken from experimental 3D tomographic reconstruction. Microsphere was irradiated from the top by linearly polarized plane wave. Simulation volume was limited by perfectly matched layers, while the size of the elementary cell was fixed at $1 \times 1 \times 1$ nm^3.

3. Results and Discussions

3.1. PLAL Synthesis of the Ag-Si Hybrids and Their Characterization

Morphology of the nanomaterial produced upon direct ns-laser ablation of monocrystalline Si wafer in the working solution containing isopropanol and $AgNO_3$ (5×10^{-4} M) for 2 h is illustrated in Figure 1b. PLAL-synthesized hybrids drop-casted onto the Si wafer generally represent spherical-shape Si particles decorated with isolated or merged Ag nano-clusters. Statistical analysis of the multiple SEM images gave an average diameter of the Ag-Si microspheres of $\approx$1.1 μm with a rather broad distribution of their size ranging from 0.7 to 1.6 μm (according to the FWHM of the Gaussian fit; Figure 1c).

Taking into account rather long 7-ns laser pulses used in the experiments, the Ag nanoclusters are expected to be produced upon thermal-induced decomposition of the $AgNO_3$ to metallic silver phase [42,43]. Once the working solution is completely transparent at 532 nm wavelength laser radiation, the photo-induced $AgNO_3$ decomposition is expected to be much less efficient for the chosen laser fluence and low concentration of the $AgNO_3$ ($<10^{-3}$ M), while no presence of the pure Ag nanoparticles was found for similar experiments performed upon irradiation of the working solution without Si wafer. Meanwhile, as the Si particles are formed and ejected into the working solution upon PLAL synthesis, these particles further act as a local laser-radiation absorbers and nano-heaters that give surface sites for preferential growth of the Ag nanoclusters. For the fixed PLAL parameters (laser wavelength, pulse duration, fluence, and irradiation time), the initial amount of the $AgNO_3$ in the working solution allows control over the amount and the geometrical shape of the formed Ag nanoclusters. More specifically, the surface density of these metallic nanoclusters gradually increases at $AgNO_3$ concentrations ranging from 10^{-5} to 10^{-2} M (Figure 1d). The separated nanoclusters typically exhibit regular shapes, according to top-view SEM images and the size ranging from 10 to 30 nm (middle panel, Figure 1d). As the density of the nanoclusters increases, they can form dimers and more complex-shape arrangements of the Si microsphere surface that are preferential for light localization and enhancement within broad spectral range. Meanwhile, at higher $AgNO_3$ concentrations (10^{-3} M), excessive formation of the Ag nanoclusters leads to their merging into larger irregular-shaped agglomerates upon laser-induced remelting. Finally, even larger content of $AgNO_3$ stimulates enormous growth and merging of Ag nanoparticles in the working solution without any preferential growth. The resulting nanomaterial mainly contained pure Ag nanoparticles with their size reaching up to 500 nm, while small amounts of Ag-Si hybrids arr observed (bottom panel, Figure 1d). Related EDX studies of the Ag-Si nanopowders produced at $AgNO_3$ concentration ranging from 10^{-4} to 10^{-3} M (that provided formation of the Ag-decorated Si microspheres) showed increase of the average Ag content from 2 to 13 wt.%.

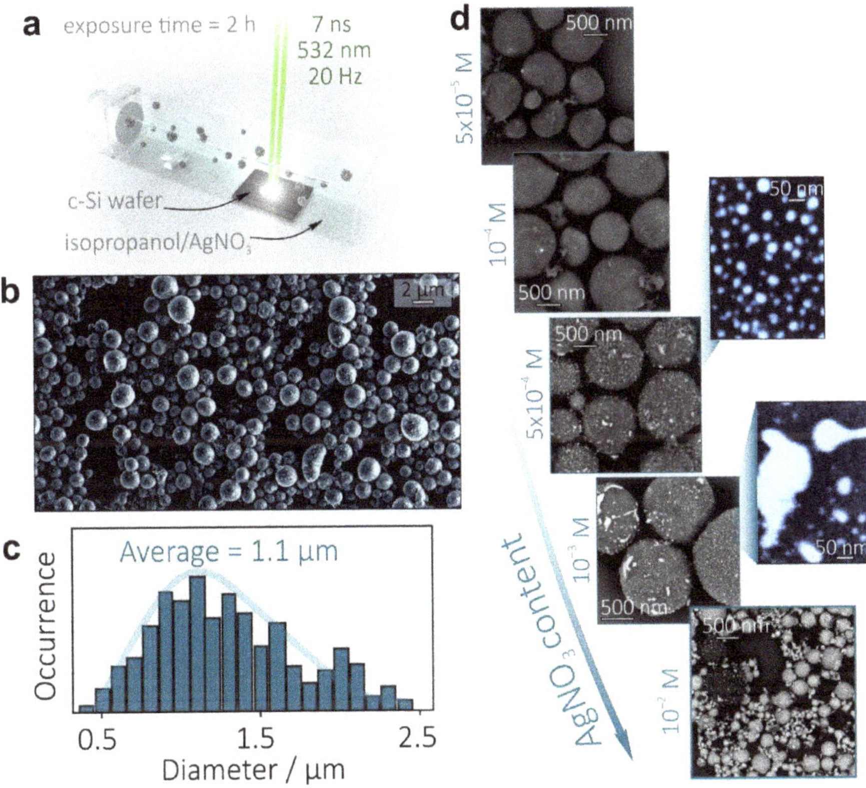

Figure 1. (**a**) Schematic of the PLAL synthesis of Ag-Si microspheres upon ns-laser ablation of the monocrystalline Si wafer placed in the isopropanol containing $AgNO_3$. (**b**) Top-view SEM image of the as-synthesized Ag-Si product dried on a Si wafer. (**c**) Size distribution of the Ag-Si microspheres. (**d**) Series of SEM images illustrating evolution of the density and geometry of the Ag nanoclusters on the surface of Si microspheres upon increase of the $AgNO_3$ content in the working solution from 5×10^{-5} to 10^{-2} M.

Deeper insight into morphology, structure, and composition of the Ag-Si hybrids produced at $AgNO_3$ content of 5×10^{-4} M was provided by combining advanced TEM imaging modalities including 3D tomographic reconstruction and high-resolution EDX elemental mapping. Figure 2a provides the STEM image of one representative cross-sectional FIB cut made through the geometric center of two neighboring 1.1-µm diameter Ag-Si hybrids. EDX analysis shows that negligibly small amount of Ag can be observed inside the bulk of the Si microsphere (bottom panel, Figure 2b). Few nm thick oxide shell was found to wrap the Si microsphere, indicating week oxidation process of the silicon upon PLAL synthesis (top panel, Figure 2b). EDX mapping also indicates presence of carbon species ($\approx$3–5 wt.%) evidently coming from decomposition of the isopropanol molecules. Appearance of pure Ag colloid agglomerating in the free space between two Ag-Si microspheres and the Si wafer can indicate non-preferential growth of the Ag nanoparticles in the working solution and/or their laser-induced removal from the Si microsphere surface by subsequent laser pulses. Further optimization of the PLAL synthesis parameters (exposure time and fluence) are to be carried out to reduce or completely remove (if required) the free Ag colloid from the resulting product. Meanwhile, such optimization goes beyond the scope of this work.

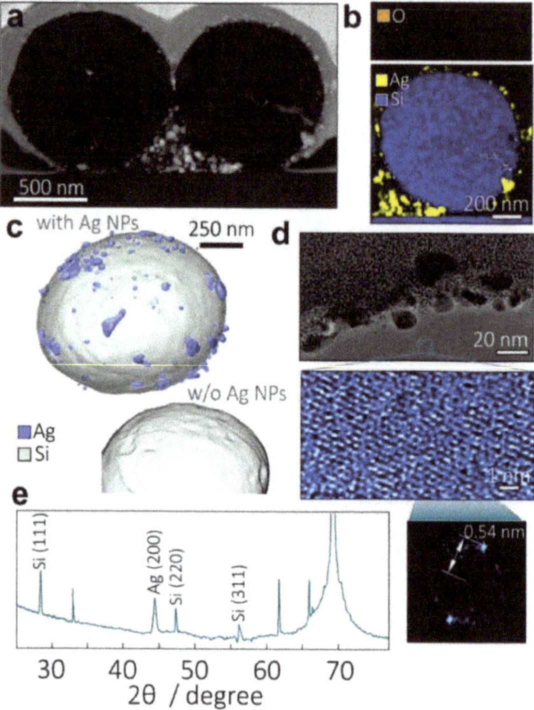

Figure 2. (**a**) STEM image of the cross-sectional central cut made through the 1.1-μm diameter Ag-Si microspheres; (**b**) EDX composition mapping of the right-most Ag-Si microsphere showing distribution of Ag and Si elements as well as an oxide shell (top panel); (**c**) 3D model of the isolated Ag-Si microsphere made through SEM tomographic reconstruction from the series of multiple images of cross-sectional cuts. Bottom panel shows the same model where Ag nanoclusters were removed to illustrate crater-like Si surface morphology; (**d**) TEM image of the Ag-Si microsphere surface (top panel) as well as HR-TEM image and its FFT showing crystalline structure of the Si core (bottom panel); (**e**) XRD pattern of the Ag-Si nanopowder. All unmarked peaks represent contribution from the underlying Si substrate.

3D tomographic model created by post-processing of multiple SEM images of the FIB cross-sectional cuts is shown in Figure 2c, providing visual perception of the Ag-Si microsphere morphology and composition. This data also suggests that certain Ag nanoclusters (both isolated and merged) are partially embedded to the Si microsphere surface (bottom panel, Figure 2c), while the Si core surface has a crater-like morphology.

The same justification can be made upon analyzing close-up TEM images of the Ag-Si microsphere surface shown in Figure 2d. HR-TEM imaging also confirms the crystalline structure of the Si core (bottom panel, Figure 2d) and its lattice constant ≈0.54 nm according to the Fast Fourier Transform analysis. XRD studies of the Ag-Si nanopowder finalize comprehensive structural and compositional characterization of the produced nanomaterial. The sharp diffraction peak at $2\Theta = 28.36°$ (see Figure 2e) corresponding to the diffraction from Si(111) planes together with less intensive peaks at 47.3° and 56.1° originate from that of Si(220) and Si(311), respectively. Despite sub-micron diameter of the obtained Ag-Si microspheres, they demonstrate rather high crystallinity with a calculated mean size of the crystallite ~180 nm in accordance with Scherrer's equation. Ag contribution is presented by a diffraction peak at 44.4° with its FWHM of 0.31°, which provides Ag crystallites mean size to be 28 nm. No signal features related to Ag silicides (either Ag_2Si or Ag_3Si) [44,45] were observed in the XRD patterns. However, related Ag(200) peak is shifted towards smaller

2Θ (larger Ag lattice constant) compared to those for bulk Ag, which can be explained by certain Ag-Si alloying upon laser ablation.

3.2. SERS Performance of the Ag-Si Hybrids

As it was mentioned in the introduction section, nanostructures made of plasmonic metals and semiconductors represent promising platform for reliable optical nanosensing based on the SERS effect. To justify the applicability of the PLAL-synthesized Ag-Si microspheres (produced at $AgNO_3$ concentrations of 5×10^{-4} M), we first functionalized them with R6G molecules (10^{-6} M alcohol solution) representing a common molecular probe that exhibits good affinity to noble metals and high Raman yield. The R6G-capped hybrids were then drop-casted onto a smooth Ag mirror in such a way to form the isolated structures on the surface that was confirmed by SEM imaging (Figure 3a; left panel). As can be seen, hybrids can also agglomerate to form a dimers or more complicated surface arrangements. However, to quantitatively assess SERS performance of PLAL-synthesized hybrids we restricted out study by the case of isolated Ag-Si microspheres.

Inherent Raman scattering of Si core (Raman band at $\approx$520 cm^{-1}) allows the exact position of Ag-Si microspheres on the surface to be easily found, aligns them with respect to the pump laser source, and relates their spatial position with the detected SERS signal from the analyte molecules. The Raman image (520 cm^{-1}) of a representative 1-µm diameter Ag-Si microsphere measured using 473 and 633 nm pump wavelengths are provided in Figure 3a. The images highlight the ability of the setup to easily resolve the isolated hybrids, Raman mapping resolution $\approx$ 1 µm and a uniform distribution of the signal with its maximum matching the center of the Ag-Si microsphere. For both pump wavelengths, we also plotted the R6G SERS images (characteristic band at 612 cm^{-1}; insets of Figure 3b) showing rather random signal distributions for different wavelengths within the microsphere area highlighted by the white ring. R6G SERS spectra averaged over the integral signal coming from the nanostructure (highlighted by white ring) are also plotted in Figure 3b, clearly showing main characteristic bands of the analyte molecules, yet with a strong contribution of the analyte photoluminescence upon its excitation at 473 nm. By comparing the averaged R6G SERS yield (total intensity of the bands at 612 and 1652 cm^{-1}) coming from the microsphere area ($\approx$3.92 $\times$ 10^4 and 5.71 $\times$ 10^4 cps per 1 mW/µm^2 for 473 and 633 nm pump wavelength, respectively) with the same signal from the R6G-capped smooth Ag mirror (4–10 cps per 1 mW/µm^2 for both wavelength), we estimated the electromagnetic contribution to the SERS enhancement factor to be at least ~10^4 for single Ag-Si microsphere. Considering additional chemical contributions coming from the interaction of the Raman probe with plasmon metal, the total SERS enhancement factor reaches 10^6 and is expected to be even larger for agglomerated microspheres.

Interestingly, the R6G SERS yield at the 633 nm pump wavelength is evidently higher than it is counter-intuitive, considering that the main contribution is expected to come from Ag nanoclusters that should exhibit stronger plasmonic response at shorter wavelengths. To clarify this feature, we carried out FDTD simulations of the local EM-field structure near the Ag-Si microsphere surface at various pump wavelengths ranging from 405 to 633 nm. 3D model of the microsphere for simulations was elaborated using the data from tomographic SEM reconstruction (Figure 2c). The main results of these simulations are summarized in Figure 3c, revealing several interesting observations. First of all, the main EM "hot spots" are expectedly related to the Ag nanoclusters, while the most intense normalized EM-field amplitude (E/E_0) is observed at 633 nm pump providing electromagnetic SERS enhancement factor~$(E/E_0)^4$ up to 10^5. Indeed, the Ag nanoclusters with the size comparable to the double skin depth should demonstrate plasmonic response in the UV spectral range. However, these clusters are strongly attached or partially embedded into the Si core, as was shown by the microscopic analysis (Figure 2c,d).

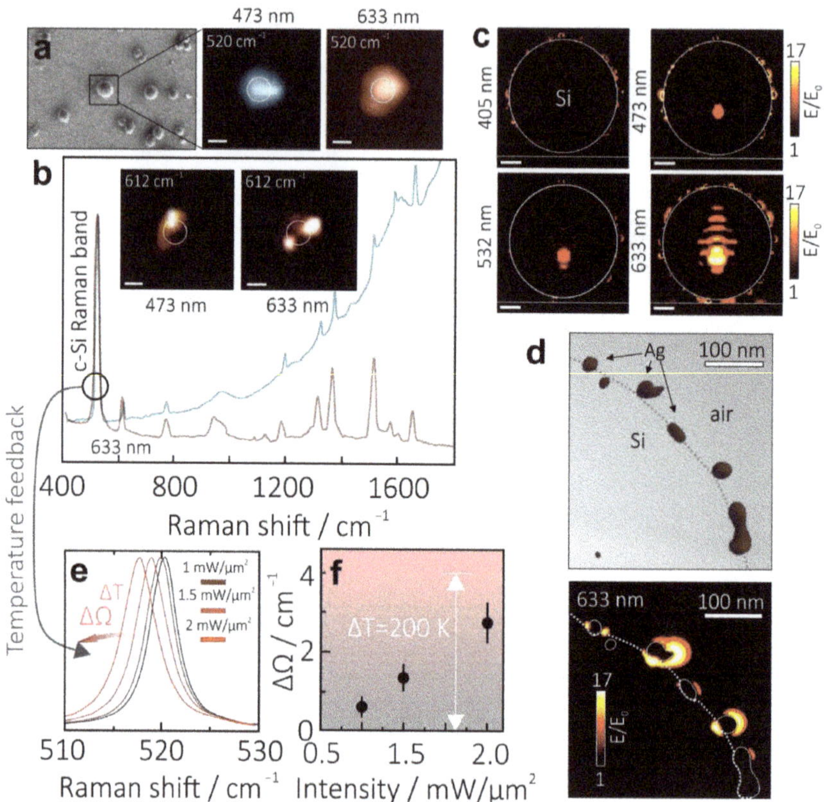

Figure 3. (**a**) Reference SEM image of the Ag-Si microspheres drop-casted onto the Si wafer as well as Raman images (at $520 \pm 2 \text{ cm}^{-1}$) of the chosen isolated nanoparticle at 473 and 633 nm laser pump wavelengths. (**b**) Representative averaged SERS spectra of the R6G capping the same Ag-Si microsphere at 473 and 633 nm laser pump wavelengths. Top insets show the distribution of the SERS signal (at $612 \pm 2 \text{ cm}^{-1}$) near the microsphere. (**c**) Normalized electric-field amplitude E/E_0 near the isolated Ag-Si microsphere on the Ag mirror upon its excitation with a linearly polarized plane wave at 405, 473, 532, and 633 nm. (**d**) Close-up distribution of E/E_0 near the isolated Ag nanoclusters on the Si surface at 633 nm pump as well as corresponding fragment of the 3D tomographic model of the Ag-Si microsphere used for modeling. (**e**) Shift of the detected c-Si Raman band at 520 cm^{-1} upon increasing the laser pump intensity (633 nm) of the Ag-Si microsphere from 0.5 to 2 mW/μm^2. (**f**) Thermally induced shift of the c-Si band $\Delta\Omega$ as a function of laser pump intensity.

Strong interaction of the plasmonic nanoparticles with high-refractive-index Si usually causes redshift and hybridization of localized plasmon resonances. Irregular elongated shape of the Ag nanoclusters and inter-particle interaction also reasonably explain redshift of the plasmonic resonances (Figure 3d). Meanwhile, the main reason that the high SERS yield at 633 nm pump is related to gradually increasing transparency of the Si material in the red part of the spectrum. This feature allows the Si core to act as an efficient Fabry–Perot cavity upon pumping from the top by the plane wave, which leads to multiple excitation cycles of the interface Ag nanoclusters by the radiation bounded within the Si core. Localization of the EM field inside the Si core as well as standing-wave-like amplitude modulations confirms this deduction. Transparent Si core also allows the pumping of the Ag nanoclusters located in the bottom hemisphere that cannot be pumped by the laser radiation incident from the top in the case of the opaque core (for example, at 405 nm pump).

Apart from the convenient ability to identify spatial position of the Ag-Si microspheres, characteristic c-Si Raman band also allows to trace the local temperature in the analyte-nanoparticle system. Analytes for SERS experiments usually represent organic or biological species with rather low thermal decomposition temperatures (i.e., R6G molecules decompose slightly above 500 K [46]). Thus, it is important to maintain the temperature at a reasonable level during SERS experiment. Upon laser-induced heating, the maximum of the c-Si Raman band blue-shifts, as it is illustrated by representative series of Raman spectra measured from the isolated Ag-Si microsphere, (produced at $AgNO_3$ concentrations of 5×10^{-4} M) pumped at various laser intensity up to 2 mW/μm^2 (Figure 3e).

The spectral shift $\Delta\Omega$ can be then recalculated to local temperature ΔT increase using the following formula [47]:

$$\Delta\Omega = A\left(1 + \frac{2}{e^{\frac{\hbar\Omega_0}{2kT}} - 1}\right) + B\left(1 + \frac{3}{e^{\frac{\hbar\Omega_0}{3kT}} - 1} + \frac{3}{\left(e^{\frac{\hbar\Omega_0}{2kT}} - 1\right)^2}\right) \tag{1}$$

where $\Omega_0 = 528$ cm^{-1}, $A = -2.96$ cm^{-1}, $B = -0.174$ cm^{-1}, $\hbar = 1.054 \cdot 10^{-34}$ J·s, $k = 1.38 \cdot 10^{-23}$ J/K. Results of systematic measurement of the heating efficiency of the isolated Ag-Si microspheres of nearly similar diameter (1.1 ± 0.15 μm) at 633 nm pump are summarized in Figure 3e, showing that pump intensity of 2 mW/μm^2 typically causes temperature ΔT increase up to $\approx$200 K. Importantly, local single-nanoparticle thermometry with an accuracy of $\pm$20 K and high spatial resolution can be realized in parallel with a SERS identification of the analytes, allowing more reliable and stable measurements [48]. For laser pump intensity slightly below 2 mW/μm^2, the Ag-Si microspheres produced at $AgNO_3$ concentrations of 5×10^{-4} M demonstrate good reproducibility of the R6G SERS yield according to the systematic measurements performed for various microspheres of nearly similar size (Figure 4a). Noteworthy, we also carried out systematic studies of the SERS yield, its reproducibility, and heating efficiency provided by the isolated Ag-Si microspheres PLAL-synthesized at larger (10^{-3} M) amount of $AgNO_3$. Despite evidently larger average SERS yield related to R6G probe, such microspheres also exhibit substantially larger deviation of the both SERS enhancement and heating efficiency limiting their application for reliable and precise measurements.

Strong and reliable SERS enhancement combined with local thermometry make Ag-Si hybrids promising for realization of advanced nanosensing platforms. To further support this statement, we applied them for SERS-based detection of plasmon-induced catalytic transformation of PATP to dimercaptoazobenzene (DMAB). Similarly to previously described experiments with R6G, Ag-Si microspheres were functionalized with PATP molecules (5×10^{-6} M, methanol solution) and drop-casted onto the Ag mirror substrate to form the isolated structures. SERS spectrum of pristine analyte contains only two Raman bands centered at 1079 and 1575 cm^{-1}, while upon its catalytic dimerization to DMAB four additional lines appear in the spectrum at 1145, 1192, 1394, and 1439 cm^{-1} [10,49,50]. According to thermogravimetric analysis, decomposition temperature of PATP is $\approx$450 K [51], thus pump intensity was set slight below 1.5 mW/μm^2 to avoid analyte degradation. Representative series of SERS spectra measured by increasing the laser exposure time of PATP-capped Ag-Si microsphere from 6 to 120 s is provided in Figure 4b showing appearance and gradual increase of the intensity of all DMAB-related Raman bands. Meanwhile, after 60 s of continuous laser exposure, the intensity of these bands starts to decay revealing complete dimerization of pristine molecules to the product and start of the photodegradation process. Mentioned dynamics is illustrated by Figure 4c, where the intensity of the brightest DMAB band at 1439 cm^{-1} (that is typically used to assess PATP-to-DMAB conversion dynamics) is plotted versus the overall laser exposure time. As can be seen, properly chosen (thermally safe) pump intensity allows the tracing of the mentioned conversion processes in detail within a reasonable time and at good SERS signal level revealing applicability of the designed single-particle platform for studying the molecular transformations under some external stimuli.

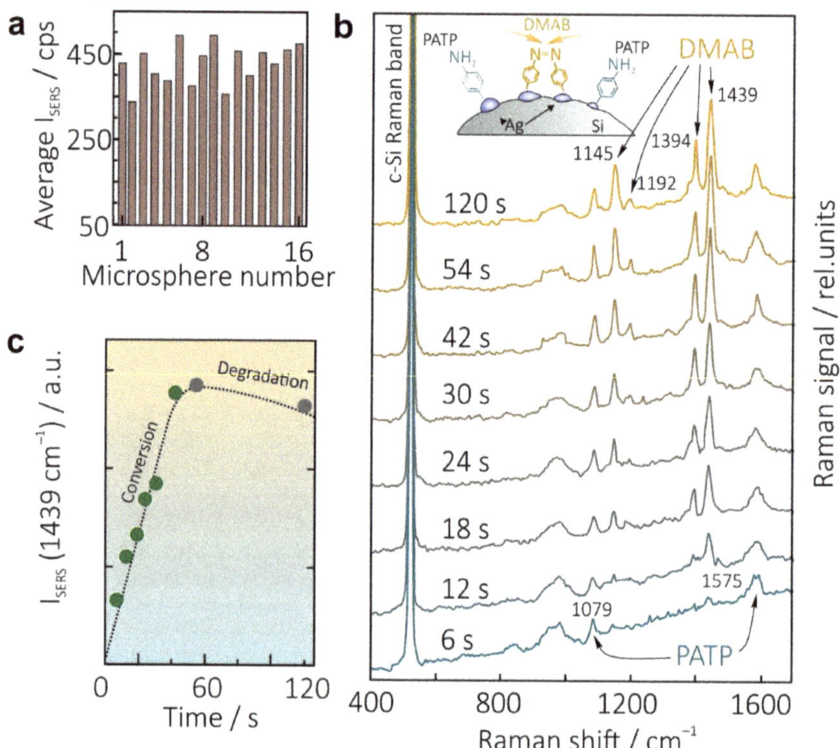

Figure 4. (**a**) R6G SERS yield (intensity of the bands at 612 and 1652 cm^{-1}) statistically averaged over measurements from 16 randomly chosen isolated Ag-Si microspheres produced at AgNO$_3$ concentrations of 5×10^{-4} M. (**b**) Series of time-resolved SERS spectra of the PATP capping the isolated Ag-Si microsphere. Total laser exposure time ranging from 6 to 120 s is indicated near each spectrum. (**c**) Evolution of the intensity of the characteristics DMAB-related Raman band at 1439 cm^{-1} as a function of laser exposure time indicating PATP-to-DMAB conversion dynamics.

4. Conclusions

To conclude, we demonstrated one-step/one-pot synthesis of Ag-decorated Si microspheres via ns-laser ablation of monocrystalline silicon in isopropanol containing AgNO$_3$. Once the ejected Si microspheres strongly absorb subsequently incident laser pulses, the preferential thermal reduction of AgNO$_3$ to metallic silver phase on the laser-heated Si microsphere surface reasonably explains their decoration process. Demonstrated submicron-scale hybrid Ag-Si particles are promising for temperature-feedback SERS bio- and chemo-sensing, while the size of the isolated microsphere ideally matches the focal spot size of the pump laser radiation, allowing reliable single-particle measurements at substantially large enhancement factor of 10^6. At the same time, the size of the Si core in the resulting Ag-Si hybrids is defined by the averaged size of the nanoparticles ejected upon ns-laser PLAL processing of the silicon surface. This means that the processing parameters can be optimized in such a way to obtain smaller averaged size of the Si nanoparticles in the suspension and thus reduce the average size of Ag-Si hybrids. In its turn, this can be achieved by decreasing the pulse duration, tuning laser wavelength that defines penetration depth of the radiation to the material, as well as by varying the type of the substrate (for example, by using Si nanowires or porous Si [52]).

Author Contributions: Conceptualization, S.G. and A.K.; methodology, V.P., A.S., S.A.K., E.M. (Eugeny Mitsai), E.M. (Evgeny Modin) and A.G.; formal analysis, S.G., A.S., S.A.K. and A.K.; resources, A.G. and E.M. (Evgeny Modin); writing—original draft preparation, A.K.; writing—review and editing, A.K., A.S. and S.A.K.; funding acquisition, A.K. All authors have read and agreed to the published version of the manuscript.

Funding: The work was supported by the Russian Science Foundation (Grant No. 21-79-00302) regarding PLAL synthesis of the nanomaterial, its characterization and SERS performance. S.A.K. acknowledges support from the Amada Foundation (grant no. AF-2019225-B3).

Institutional Review Board Statement: Not applicable.

Informed Consent Statement: Not applicable.

Data Availability Statement: Data will be made available on request.

Conflicts of Interest: The authors declare no conflict of interest.

References

1. Maier, S.A. *Plasmonics: Fundamentals and Applications*; Springer: New York, NY, USA, 2007; Volume 1, p. 245.
2. Kuznetsov, A.I.; Miroshnichenko, A.E.; Brongersma, M.L.; Kivshar, Y.S.; Luk'yanchuk, B. Optically resonant dielectric nanostructures. *Science* **2016**, *354*, aag2472. [CrossRef] [PubMed]
3. Jiang, R.; Li, B.; Fang, C.; Wang, J. Metal/semiconductor hybrid nanostructures for plasmon-enhanced applications. *Adv. Mater.* **2014**, *26*, 5274–5309. [CrossRef] [PubMed]
4. Fusco, Z.; Rahmani, M.; Tran-Phu, T.; Ricci, C.; Kiy, A.; Kluth, P.; Della Gaspera, E.; Motta, N.; Neshev, D.; Tricoli, A.S.A. Photonic Fractal Metamaterials: A Metal–Semiconductor Platform with Enhanced Volatile-Compound Sensing Performance. *Adv. Mater.* **2020**, *32*, 2002471. [CrossRef] [PubMed]
5. Atwater, H.A.; Polman, A. Plasmonics for improved photovoltaic devices. *Nat. Mater.* **2010**, *9*, 205–213. [CrossRef]
6. Li, W.; Valentine, J.G. Harvesting the loss: Surface plasmon-based hot electron photodetection. *Nanophotonics* **2017**, *6*, 177–191. [CrossRef]
7. Xiang, J.; Jiang, S.; Chen, J.; Li, J.; Dai, Q.; Zhang, C.; Xu, Y.; Tie, S.; Lan, S. Hot-electron intraband luminescence from GaAs nanospheres mediated by magnetic dipole resonances. *Nano Lett.* **2017**, *17*, 4853–4859. [CrossRef]
8. Manuel, A.P.; Shankar, K. Hot electrons in TiO_2–noble metal nano-heterojunctions: Fundamental science and applications in photocatalysis. *Nanomaterials* **2021**, *11*, 1249. [CrossRef]
9. Sharma, B.; Frontiera, R.R.; Henry, A.I.; Ringe, E.; Van Duyne, R.P. SERS: Materials, applications, and the future. *Mater. Today* **2012**, *15*, 16–25. [CrossRef]
10. Mitsai, E.; Kuchmizhak, A.; Pustovalov, E.; Sergeev, A.; Mironenko, A.; Bratskaya, S.; Linklater, D.P.; Balčytis, A.; Ivanova, E.; Juodkazis, S. Chemically non-perturbing SERS detection of a catalytic reaction with black silicon. *Nanoscale* **2018**, *10*, 9780–9787. [CrossRef]
11. Dmitriev, P.A.; Baranov, D.G.; Milichko, V.A.; Makarov, S.V.; Mukhin, I.S.; Samusev, A.K.; Krasnok, A.E.; Belov, P.A.; Kivshar, Y.S. Resonant Raman scattering from silicon nanoparticles enhanced by magnetic response. *Nanoscale* **2016**, *8*, 9721–9726. [CrossRef]
12. Zograf, G.P.; Petrov, M.I.; Zuev, D.A.; Dmitriev, P.A.; Milichko, V.A.; Makarov, S.V.; Belov, P.A. Resonant nonplasmonic nanoparticles for efficient temperature-feedback optical heating. *Nano Lett.* **2017**, *17*, 2945–2952. [CrossRef] [PubMed]
13. Zhang, D.; Li, Z.; Liang, C. Diverse nanomaterials synthesized by laser ablation of pure metals in liquids. *Sci. China Phys. Mech. Astron.* **2022**, *65*, 1–35. [CrossRef]
14. Amendola, V.; Amans, D.; Ishikawa, Y.; Koshizaki, N.; Scirè, S.; Compagnini, G.; Reichenberger, S.; Barcikowski, S. Room-temperature laser synthesis in liquid of oxide, metal-oxide core-shells, and doped oxide nanoparticles. *Chem. –A Eur. J.* **2020**, *26*, 9206–9242. [CrossRef] [PubMed]
15. Zhang, D.; Gokce, B.; Barcikowski, S. Laser synthesis and processing of colloids: Fundamentals and applications. *Chem. Rev.* **2017**, *117*, 3990–4103. [CrossRef]
16. Siuzdak, K.; Sawczak, M.; Klein, M.; Nowaczyk, G.; Jurga, S.; Cenian, A. Preparation of platinum modified titanium dioxide nanoparticles with the use of laser ablation in water. *Phys. Chem. Chem. Phys.* **2014**, *16*, 15199–15206. [CrossRef]
17. Hamad, A.; Li, L.; Liu, Z.; Zhong, X.L.; Wang, T. Picosecond laser generation of Ag–TiO2 nanoparticles with reduced energy gap by ablation in ice water and their antibacterial activities. *Appl. Phys. A* **2015**, *119*, 1387–1396. [CrossRef]
18. Stadnichenko, A.; Svintsitskiy, D.; Kibis, L.; Fedorova, E.; Stonkus, O.; Slavinskaya, E.; Lapin, I.; Fakhrutdinova, E.; Svetlichnyi, V.; Romanenko, A.; et al. Influence of titania synthesized by pulsed laser ablation on the state of platinum during ammonia oxidation. *Appl. Sci.* **2020**, *10*, 4699. [CrossRef]
19. Gurbatov, S.O.; Modin, E.; Puzikov, V.; Tonkaev, P.; Storozhenko, D.; Sergeev, A.; Mintcheva, N.; Yamaguchi, S.; Tarasenka, N.N.; Chuvilin, A.; et al. Black Au-decorated TiO_2 produced via laser ablation in liquid. *ACS Appl. Mater. Interfaces* **2021**, *13*, 6522–6531. [CrossRef]

20. Mintcheva, N.; Srinivasan, P.; Rayappan JB, B.; Kuchmizhak, A.A.; Gurbatov, S.; Kulinich, S.A. Room-temperature gas sensing of laser-modified anatase TiO$_2$ decorated with Au nanoparticles. *Appl. Surf. Sci.* **2020**, *507*, 145169. [CrossRef]

21. Gurbatov, S.; Puzikov, V.; Storozhenko, D.; Modin, E.; Mitsai, E.; Cherepakhin, A.; Shevlyagin, A.; Gerasimenko, A.V.; Kulinich, S.A.; Kuchmizhak, A. Hybrid Au-Si microspheres produced by laser ablation in liquid (LAL) for temperature-feedback optical nano-sensing and anti-counterfeit labeling. *arXiv* **2022**, arXiv:2204.05124.

22. Larin, A.; Nomin´e, A.; Ageev, E.; Ghanbaja, J.; Kolotova, L.; Starikov, S.; Bruy´ere, S.; Belmonte, T.; Makarov, S.; Zuev, D. Plasmonic nanosponges filled with silicon for enhanced white light emission. *Nanoscale* **2020**, *12*, 1013–1021. [CrossRef] [PubMed]

23. Liu, P.; Chen, H.; Wang, H.; Yan, J.; Lin, Z.; Yang, G. Fabrication of Si/Au core/shell nanoplasmonic structures with ultrasensitive surface-enhanced Raman scattering for monolayer molecule detection. *J. Phys. Chem. C* **2015**, *119*, 1234–1246. [CrossRef]

24. Ryabchikov, Y.V. Facile laser synthesis of multimodal composite silicon/gold nanoparticles with variable chemical composition. *J. Nanoparticle Res.* **2019**, *21*, 1–10. [CrossRef]

25. Kutrovskaya, S.; Arakelian, S.; Kucherik, A.; Osipov, A.; Evlyukhin, A.; Kavokin, A. The synthesis of hybrid gold-silicon nano particles in a liquid. *Sci. Rep.* **2017**, *7*, 10284. [CrossRef]

26. Al-Kattan, A.; Tselikov, G.; Metwally, K.; Popov, A.A.; Mensah, S.; Kabashin, A.V. Laser Ablation-Assisted Synthesis of Plasmonic Si@ Au Core-Satellite Nanocomposites for Biomedical Applications. *Nanomaterials* **2021**, *11*, 592. [CrossRef]

27. Zeng, H.; Du, X.W.; Singh, S.C.; Kulinich, S.A.; Yang, S.; He, J.; Cai, W. Nanomaterials via laser ablation/irradiation in liquid: A review. *Adv. Funct. Mater.* **2012**, *22*, 1333–1353. [CrossRef]

28. Pini, F.; Pilot, R.; Ischia, G.; Agnoli, S.; Amendola, V. Au–Ag Alloy Nanocorals with Optimal Broadband Absorption for Sunlight-Driven Thermoplasmonic Applications. *ACS Appl. Mater. Interfaces* **2022**, *14*, 28924–28935. [CrossRef]

29. Tarasenka, N.; Shustava, E.; Butsen, A.; Kuchmizhak, A.A.; Pashayan, S.; Kulinich, S.A.; Tarasenko, N. Laser-assisted fabrication and modification of copper and zinc oxide nanostructures in liquids for photovoltaic applications. *Appl. Surf. Sci.* **2021**, *554*, 149570. [CrossRef]

30. Nastulyavichus, A.A.; Saraeva, I.N.; Rudenko, A.A.; Khmelnitskii, R.A.; Shakhmin, A.L.; Kirilenko, D.A.; Brunkov, P.N.; Melnik, N.N.; Smirnov, N.A.; Ionin, A.A.; et al. Multifunctional Sulfur-Hyperdoped Silicon Nanoparticles with Engineered Mid-Infrared Sulfur-Impurity and Free-Carrier Absorption. *Part. Part. Syst. Charact.* **2020**, *37*, 2000010. [CrossRef]

31. Seifikar, F.; Azizian, S.; Eslamipanah, M.; Jaleh, B. One step synthesis of stable nanofluids of Cu, Ag, Au, Ni, Pd, and Pt in PEG using laser ablation in liquids method and study of their capability in solar-thermal conversion. *Sol. Energy* **2022**, *246*, 74–88. [CrossRef]

32. Zhang, D.; Gökce, B.; Notthoff, C.; Barcikowski, S. Layered seed-growth of AgGe football-like microspheres via precursor-free picosecond laser synthesis in water. *Sci. Rep.* **2015**, *5*, 13661. [CrossRef] [PubMed]

33. Shankar, P.; Ishak, M.H.; Padarti, J.K.; Mintcheva, N.; Iwamori, S.; Gurbatov, S.O.; Lee, J.H.; Kulinich, S.A. ZnO@ graphene oxide core@ shell nanoparticles prepared via one-pot approach based on laser ablation in water. *Appl. Surf. Sci.* **2020**, *531*, 147365. [CrossRef]

34. Rehbock, C.; Merk, V.; Gamrad, L.; Streubel, R.; Barcikowski, S. Size control of laser-fabricated surfactant-free gold nanoparticles with highly diluted electrolytes and their subsequent bioconjugation. *Phys. Chem. Chem. Phys.* **2013**, *15*, 3057–3067. [CrossRef]

35. Torresan, V.; Forrer, D.; Guadagnini, A.; Badocco, D.; Pastore, P.; Casarin, M.; Selloni, A.; Coral, D.F.C.; Ceolin, M.; Fernandez van Raap, M.B.; et al. 4D multimodal nanomedicines made of Nonequilibrium Au–Fe alloy nanoparticles. *Acs Nano* **2020**, *14*, 12840–12853. [CrossRef]

36. Alexander, D.T.; Forrer, D.; Rossi, E.; Lidorikis, E.; Agnoli, S.; Bernasconi, G.D.; Butet, J.; Martin, O.J.; Amendola, V. Electronic structure-dependent surface plasmon resonance in single Au–Fe nanoalloys. *Nano Lett.* **2019**, *19*, 5754–5761. [CrossRef]

37. Vassalini, I.; Borgese, L.; Mariz, M.; Polizzi, S.; Aquilanti, G.; Ghigna, P.; Sartorel, A.; Amendola, V.; Alessandri, I. Enhanced electrocatalytic oxygen evolution in Au–Fe nanoalloys. *Angew. Chem. Int. Ed.* **2017**, *56*, 6589–6593. [CrossRef] [PubMed]

38. Dolgaev, S.I.; Simakin, A.V.; Voronov, V.V.; Shafeev, G.A.; Bozon-Verduraz, F. Nanoparticles produced by laser ablation of solids in liquid environment. *Appl. Surf. Sci.* **2002**, *186*, 546–551. [CrossRef]

39. Amendola, V.; Meneghetti, M. Laser ablation synthesis in solution and size manipulation of noble metal nanoparticles. *Phys. Chem. Chem. Phys.* **2009**, *11*, 3805–3821. [CrossRef] [PubMed]

40. Saraeva, I.N.; Van Luong, N.; Kudryashov, S.I.; Rudenko, A.A.; Khmelnitskiy, R.A.; Shakhmin, A.L.; Kharin, A.Y.; Ionin, A.A.; Zayarny, D.A.; Van Duong, P.; et al. Laser synthesis of colloidal Si@ Au and Si@ Ag nanoparticles in water via plasma assisted reduction. *J. Photochem. Photobiol. A Chem.* **2018**, *360*, 125–131. [CrossRef]

41. Palik, E.D. (Ed.) *Handbook of Optical Constants of Solids*; Academic Press: Cambridge, MA, USA, 1998; Volume 3.

42. Fox, B.S.; Beyer, M.K.; Bondybey, V.E. Coordination chemistry of silver cations. *J. Am. Chem. Soc.* **2002**, *124*, 13613. [CrossRef]

43. Borodaenko, Y.; Syubaev, S.; Khairullina, E.; Tumkin, I.; Gurbatov, S.; Mironenko, A.; Mitsai, E.; Zhizhchenko, A.; Modin, E.; Gurevich, E.L.; et al. On-demand Plasmon Nanoparticle-Embedded Laser-Induced Periodic Surface Structures (LIPSSs) on Silicon for Optical Nanosensing. *Adv. Opt. Mater.* **2022**, *10*, 2201094. [CrossRef]

44. Liu, B.; Xu, G.; Jin, C.; Yang, X.; Kong, K.; Ouyang, P.; Zou, Y.; Yang, W.; Yue, Z.; Li, X.; et al. The Si/Ag$_2$Si/Ag particles with the enhanced mechanical contact as anode material for lithium ion batteries. *Mater. Lett.* **2020**, *280*, 128536. [CrossRef]

45. Leedy, K.D.; Rigsbee, J.M. Microstructure of radio frequency sputtered Ag$_{1-x}$Si$_x$ alloys. *J. Vac. Sci. Technol. A Vac. Surf. Film.* **1996**, *14*, 2202–2206. [CrossRef]

46. Smitha, V.S.; Manjumol, K.A.; Ghosh, S.; Brahmakumar, M.; Pavithran, C.; Perumal, P.; Warrier, K.G. Rhodamine 6g intercalated montmorillonite nanopigments–polyethylene composites: Facile synthesis and ultravioletstability study. *J. Am. Ceram. Soc.* **2011**, *94*, 1731–1736. [CrossRef]
47. Balkanski, M.; Wallis, R.F.; Haro, E. Anharmonic effects in light scattering due to optical phonons in silicon. *Phys. Rev.* **1983**, *28*, 1928. [CrossRef]
48. Aouassa, M.; Mitsai, E.; Syubaev, S.; Pavlov, D.; Zhizhchenko, A.; Jadli, I.; Hassayoun, L.; Zograf, G.; Makarov, S.; Kuchmizhak, A. Temperature-feedback direct laser reshaping of silicon nanostructures. *Appl. Phys. Lett.* **2017**, *111*, 243103. [CrossRef]
49. Huang, Y.F.; Zhu, H.P.; Liu, G.K.; Wu, D.Y.; Ren, B.; Tian, Z.Q. When the signal is not from the original molecule to be detected: Chemical transformation of para-aminothiophenol on Ag during the SERS measurement. *J. Am. Chem. Soc.* **2010**, *132*, 9244–9246. [CrossRef] [PubMed]
50. Sun, M.; Xu, H. A novel application of plasmonics: Plasmon-driven surface-catalyzed reactions. *Small* **2012**, *8*, 2777–2786. [CrossRef]
51. Mitsai, E.; Naffouti, M.; David, T.; Abbarchi, M.; Hassayoun, L.; Storozhenko, D.; Mironenko, A.; Bratskaya, S.; Juodkazis, S.; Makarov, S.; et al. Si 1 − x Ge x nanoantennas with a tailored raman response and light-to-heat conversion for advanced sensing applications. *Nanoscale* **2019**, *11*, 11634–11641. [CrossRef]
52. Zabotnov, S.V.; Skobelkina, A.V.; Sergeeva, E.A.; Kurakina, D.A.; Khilov, A.V.; Kashaev, F.V.; Kaminskaya, T.P.; Presnov, D.E.; Agrba, P.D.; Shuleiko, D.V.; et al. Nanoparticles produced via laser ablation of porous silicon and silicon nanowires for optical bioimaging. *Sensors* **2020**, *20*, 4874. [CrossRef]

 *materials*

Article

Cu Patterning Using Femtosecond Laser Reductive Sintering of CuO Nanoparticles under Inert Gas Injection

Mizue Mizoshiri * and Kyohei Yoshidomi

Department of Mechanical Engineering, Nagaoka University of Technology, Niigata 940-2188, Japan; s193097@stn.nagaokaut.ac.jp
* Correspondence: mizoshiri@mech.nagaokaut.ac.jp; Tel.: +81-258-47-9765

Abstract: In this paper, we report the effect of inert gas injection on Cu patterning generated by femtosecond laser reductive sintering of CuO nanoparticles (NPs). Femtosecond laser reductive sintering for metal patterning has been restricted to metal and metal-oxide composite materials. By irradiating CuO-nanoparticle paste with femtosecond laser pulses under inert gas injection, we intended to reduce the generation of metal oxides in the formed patterns. In an experimental evaluation, the X-ray diffraction peaks corresponding to copper oxides, such as CuO and Cu_2O, were much smaller under N_2 and Ar gas injections than under air injection. Increasing the injection rates of both gases increased the reduction degree of the X-ray diffraction peaks of the CuO NPs, but excessively high injection rates ($\geq$100 mL/min) significantly decreased the surface density of the patterns. These results qualitatively agreed with the ratio of sintered/melted area. The femtosecond laser reductive sintering under inert gas injection achieved a vacuum-free direct writing of metal patterns.

Keywords: Cu pattern; CuO nanoparticle ink; femtosecond laser reductive sintering; inert gas injection

Citation: Mizoshiri, M.; Yoshidomi, K. Cu Patterning Using Femtosecond Laser Reductive Sintering of CuO Nanoparticles under Inert Gas Injection. *Materials* **2021**, *14*, 3285. https://doi.org/10.3390/ma14123285

Academic Editor: Alina A. Manshina

Received: 14 May 2021
Accepted: 7 June 2021
Published: 14 June 2021

Publisher's Note: MDPI stays neutral with regard to jurisdictional claims in published maps and institutional affiliations.

1. Introduction

Direct laser writing of metals is a promising technique for wiring and fabricating electric devices because it achieves both metallization and formation of the desired patterns. In laser direct writing, various inks based on metal nanoparticles (NPs), metal complexes, metal–organic decompositions, and metal-oxide NPs are coated on substrates, and then selectively metallized by irradiation with focused laser light. Inks that comprised noble metal (such as Au and Ag) NPs are commercially available and achieve highly electroconductive patterns in ambient atmospheres [1,2].

Meanwhile, Cu patterning is a potential candidate for a low-cost printing [3]. Presently, Cu nanoparticle inks are selectively sintered by continuous-wave (CW) lasers or by nanosecond, picosecond, and femtosecond pulsed lasers [4–7]. However, unlike noble metal patterns, Cu patterns are not easily formed in air because the Cu NPs in the inks and fabricated patterns are easily oxidized by the atmospheric oxygen. Therefore, the development of various inks based on Cu-organic decomposition, Cu complexes, and Cu-oxide NPs are carried out. Cu patterns are formed through laser light-induced thermochemical reduction. For example, Cu-organic decomposition and Cu-complex inks coated on substrates are selectively reduced by irradiating them with a CW or pulsed laser, which precipitates the Cu [8–10]. In Cu-organic decomposition ink, Cu patterns are formed from Cu(II) formate by a thermal reaction. The gases generated in the decomposition process (CO_2 and H_2) provide a reductive atmosphere that prevents the oxidation of the Cu patterns [10]. A glyoxylic acid Cu complex ink for Cu laser direct writing in air has also been developed. On glass substrates coated with this ink, Cu patterns without significant oxidation are formed by a CO_2 laser-induced thermochemical reaction [8,9]. Other promising candidates for Cu pattering are Cu-oxide NP inks that comprised Cu-oxide NPs, a reductant, and a dispersant. When mixed with a reductant and a dispersant, Cu_2O NPs are reduced to Cu by a laser-induced thermochemical reaction (where the laser is CW, nanosecond-pulsed, or femtosecond pulsed [11–13]). In addition, CuO nanoparticles are also reduced and sintered by various lasers, such as continuous-wave, nanosecond, and femtosecond pulse lasers with green and near-infrared wavelength [3,14–20]. The irradiated laser light reduces and sinters the Cu-oxide NPs while generating various gases such as CO, CO_2, and H_2O. The advantage of the use of Cu-oxide NP inks is their small volume change after irradiation with laser light. The volume remains more-or-less constant because the Cu in the inks has a higher volumetric composition than the Cu precipitated from Cu-organic decomposition and Cu-complex inks. Thermochemical reduction of CuO NPs using CW and nanosecond lasers generates Cu patterns with no significant oxidation [3,14]. The advantage of the femtosecond laser reductive sintering is that the CuO NPs were reductive-sintered under low pulse energy [15,17,21] by comparing to the use of CW and nanosecond pulsed lasers [3,14]. By contrast, reductive sintering with femtosecond lasers selectively form Cu-and Cu_2O-rich patterns [15,17,21]. In the X-ray diffraction (XRD) spectra of Cu-rich patterns fabricated under femtosecond laser pulses, significant peaks corresponding to Cu_2O and CuO are found, whereas the Cu patterns obtained by irradiation of CW and nanosecond lasers are almost single-phase Cu. These results typify femtosecond laser reductive sintering of Cu patterning on glass and polyimide film substrates and are commonly reported [15,17,21]. As the oxidation increases with decreasing laser scanning speed [15,17], we considered that Cu becomes re-oxidized by atmospheric oxygen.

In the present study, we investigated the effect of an atmospheric gas injection on reductive sintering of CuO NPs under femtosecond laser pulses. The patterns fabricated under various gases (air, and inert gases of N_2, and Ar generally used in 3D printing and additive manufacturing) were examined for their crystal structures, surface oxidation, and morphology, and their electric resistances were evaluated.

2. Experimental Methods

2.1. Cu Patterning Based on Femtosecond Laser Reductive Sintering of CuO NPs under a Controlled Atmosphere

Figure 1 shows the proposed for Cu-patterning process based on femtosecond laser reductive sintering. First, a CuO-NP ink was prepared by using a previously reported process [14,15,17]. CuO NPs (<50 nm diameter, Sigma Aldrich, St. Louis, MS, USA) were dispersed in a mixed solution of polyvinylpyrrolidone (PVP, M_w ~10,000, Sigma Aldrich, MS, USA) and ethylene glycol (EG, Sigma Aldrich, MS, USA), and the resulting ink was spin-coated on glass substrates (thickness of ~1 mm). Before spin coating, the substrates were surface-treated with ultraviolet light for 1 min in air to improve their wettability (FLAT EXCIMER EX-mini, Hamamatsu Photonics, Hamamatsu, Japan). The surface of the glass substrates was changed to be hydrophilic by irradiating ultraviolet light which generated ozone from oxygen in air [22]. The CuO-NP ink film coated on the glass was irradiated with near-infrared femtosecond laser pulses with a wavelength, pulse duration, and repetition frequency of 780 nm, 120 fs, and 80 MHz, respectively, (Toptica, FemtoFiber pro NIR, TOPTICA Photonics AG, Munich, Germany). The laser pulses were focused through an objective lens with a numerical aperture of 0.45. The laser irradiation process was conducted under a controlled atmosphere of air, Ar gas, or N_2 gas injection. The injection rates of the gases were verified using a flow meter. Finally, the non-sintered NPs in the ink were removed by sequentially rinsing the sample in EG and followed by in ethanol.

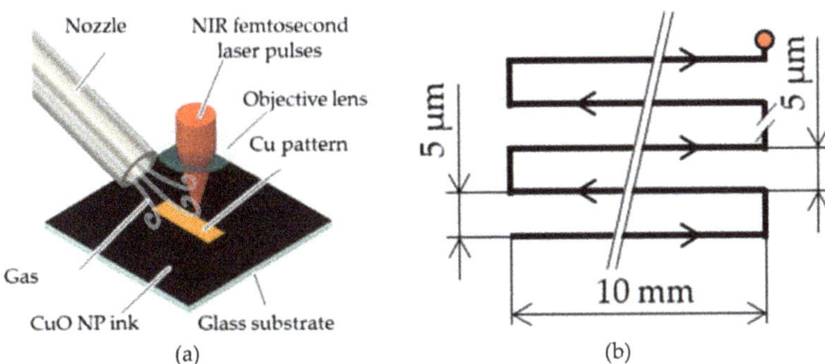

Figure 1. Schematic illustration of Cu patterning under a controlled atmosphere: (**a**) laser patterning and (**b**) scanning route of the focal spot over the paste.

2.2. Evaluation Methods

The crystal structures of the Cu patterns were examined by XRD analysis using Cu-Kα radiation with a scan speed of 5°/min. The line width was measured by using an optical microscope (Keyence, VE-8800, Osaka, Japan). The morphology of the patterns was also evaluated by using the same optical microscope and a scanning electron microscope (SEM) (HITACHI High Technology Corporation, FlexSEM 1000 II, Tokyo, Japan). The atomic composition and surface oxidation of the patterns were investigated using an energy-dispersive X-ray spectroscopy (EDS, Oxford Instruments plc, Abingdon-on-Thames, United Kingdom). Additionally, a laser Raman spectrometer (JASCO, NRS-7200, Tokyo, Japan) with a laser wavelength of 532 nm, respectively. The resistances of the patterns were measured using a multimeter (Keysight Technology, Truevolt series 34465A, Santa Rosa, CA, USA).

3. Results

3.1. Effect of Air Injection on the Pattern Uniformity

We first investigated the effect of gas injection on the pattern uniformity. Figure 2a is a photograph of the patterns fabricated with and without air injection using compressed air with controlling the flow rate, and Figure 2b shows the Raman spectra of the patterns. The patterns were fabricated at a scan speed and pulse energy of 5 mm/s and 0.74 nJ, respectively. The Raman spectroscopy was performed at measurement points No. 1, No. 2, and No. 3, in Figure 2a. In the patterns written at the starting area (No. 1), the CuO NPs were highly reduced and significant oxidation-free regardless of air injection (absent or present). By contrast, at points No.2 and No.3, copper-oxidized Cu_2O and Cu species, respectively, were found in the absence of air injection, but significant oxidation was not observed under air injection. These results suggest that air injection improves the uniformity of the beginning and end of the written area. The thermochemical reduction of CuO NPs to Cu creates water vapor that coats the surface of the objective lens, decreasing the laser pulse energy irradiated on the CuO NP paste and hence the degree of the reduction to Cu [14].

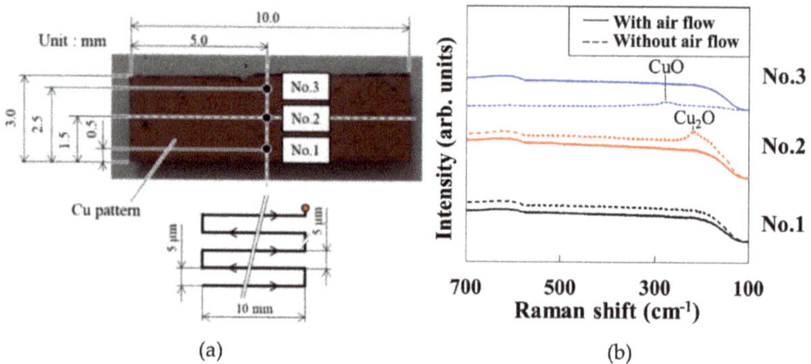

(a) (b)

Figure 2. (**a**) Photograph of the pattern fabricated under an air injection at a scan speed and pulse energy of 5 mm/s and 0.74 nJ, respectively; (**b**) Raman spectra at points No. 1, No. 2, and No. 3 of the patterns fabricated in the presence and absence of air injection.

3.2. Effect of Scan Speed on the Crystal Structures of the Patterns

We next investigated the effect of scan speed on the crystal structures. Figure 3a shows the XRD spectra of the patterns fabricated at different scan speeds. Intense peaks corresponding to Cu were observed in all spectra. Small peaks corresponding to Cu_2O and CuO were also observed. Figure 3b plots the degree of reduction as the intensity ratio of Cu(111)/CuO(111) versus the scan speed. The degree of reduction was maximized at 8 mm/s and was relatively small at higher and lower scan speeds. These results suggest low reduction and the re-oxidation of the patterns at high and low scan speeds, respectively.

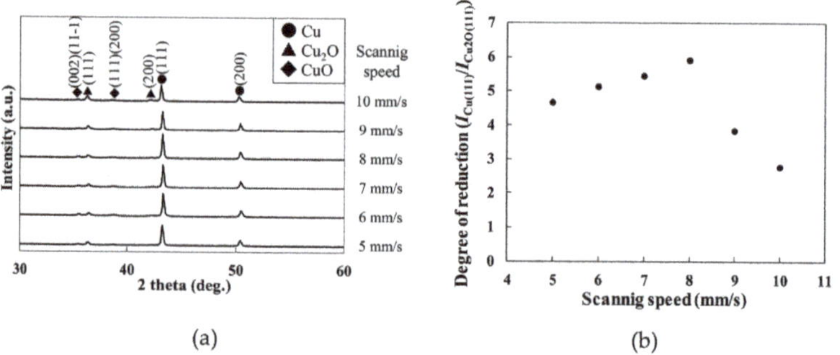

(a) (b)

Figure 3. (**a**) XRD spectra of the patterns fabricated at various scan speeds under an air injection; (**b**) degree of reduction at various scan speeds.

3.3. Crystal Structures of the Patterns Fabricated under Various Gas Injections

This subsection investigates the crystal structures of the patterns fabricated under N_2 and Ar gas injections. Figure 4a,b show the XRD spectra of the patterns fabricated under N_2 and Ar gas injections, respectively. The XRD peaks corresponding to CuO and Cu_2O decreased in both XRD spectra of both patterns. Figure 4c,d plot the degree of reductions, reported as the XRD intensity ratio $I_{Cu(111)}/I_{CuO(111)}$ and $I_{Cu(111)}/I_{Cu2O(111)}$ of the patterns formed under various gas injection rates. The Cu generation in the patterns increased with gas injection rate. The degree of the reduction was not affected by the types of gasses. This result suggests that both gases removed O_2 gas from atmospheric air during sintering and melting of the CuO NPs.

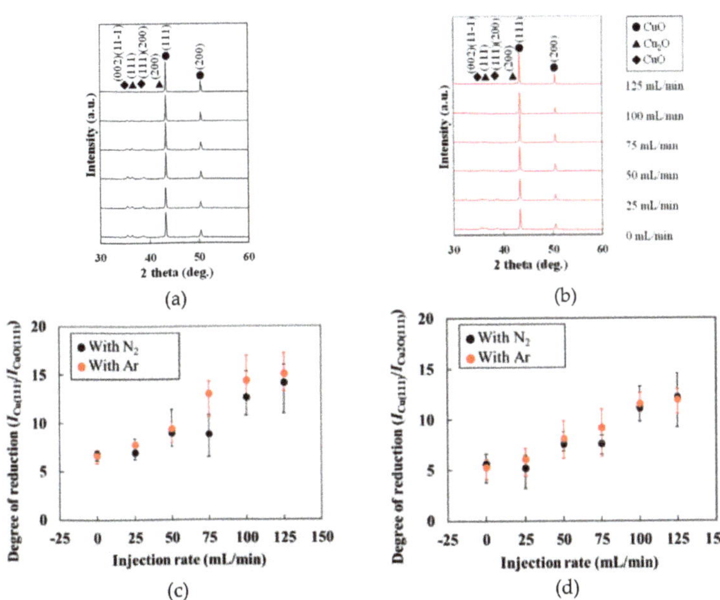

Figure 4. XRD spectra of the patterns fabricated under (**a**) N_2 gas and (**b**) Ar gas; degrees of reduction, (**c**) Cu/CuO and (**d**) Cu/Cu_2O of the patterns fabricated under gases with various injection rates.

3.4. Surface Morphology

The surface morphologies of the patterns fabricated under N_2 and Ar gases at various injection rates are revealed in the optical images (Figure 5). The copper-like brown and pore-like white area were observed under the injection rate of ≥ 100 mL/min. Figure 6 shows the SEM image EDS mappings of the brown-colored and white-colored areas of the pattern fabricated under the Ar gas injection rate of 125 mL/min, revealing that the brown-colored area was well-sintered and white-colored area was the bare glass substrates because the white-colored area was not Cu- and O-rich, established through EDS mapping, corresponding to glasses. The droplets were generated in the melted area. Table 1 lists the surface densities of the fabricated patterns. The Cu metal luster of the patterns were observed in all the optical images. The density decreased with the increasing injection rate of both gases. The patterns formed under high injection rates (100 and 125 mL/min) exhibited many droplets that apparently formed by the melting and cooling of the materials. These droplets were considered to markedly decrease the density of the patterns from those formed at lower injection rates (0–75 mL/min).

Table 1. The ratio of sintered/melted area of the patterns fabricated under N_2 and Ar gases at different injection rates.

Injection Rate (mL/min)	0	25	50	75	100	125
N_2 gas	95.1%	94.5%	94.0%	93.2%	81.8%	68.8%
Ar gas	95.9%	96.4%	96.6%	95.7%	84.9%	76.9%

Figure 7 shows the concentration ratio of the patterns in sintered and melted area. The concentration of Cu in the sintered area was approximately 70% which was not affected by the kinds of gases and their injection rate. The concentration of oxygen decreased with increasing the injection rate, that of carbon increased with an increase in both the sintered and melted area. These results agreed with the degree of reductions, Cu/CuO and Cu/Cu_2O shown in Figure 4.

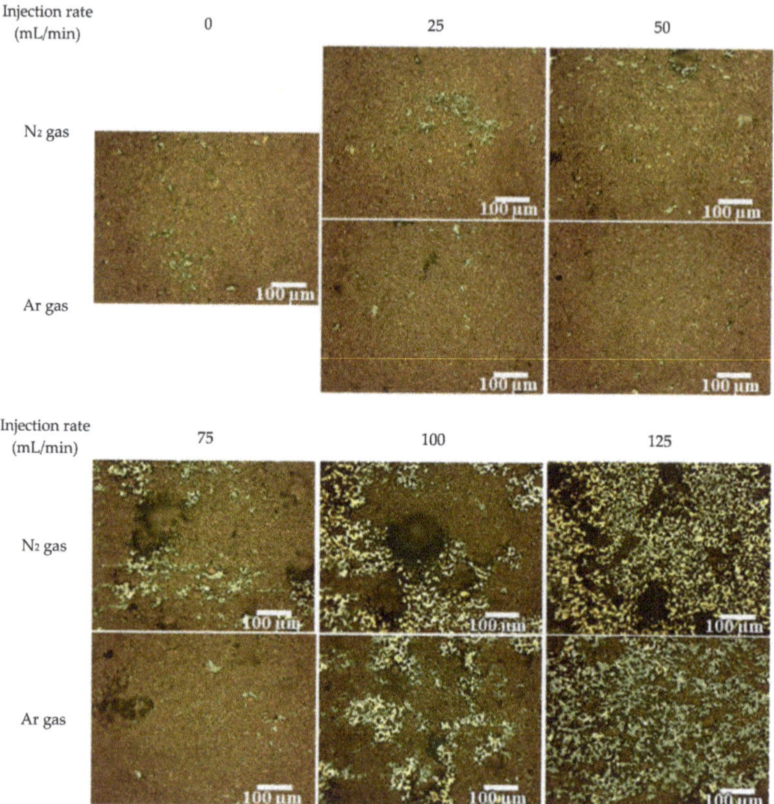

Figure 5. Optical microscope images of the patterns fabricated under various N$_2$ and Ar gases at different injection rates.

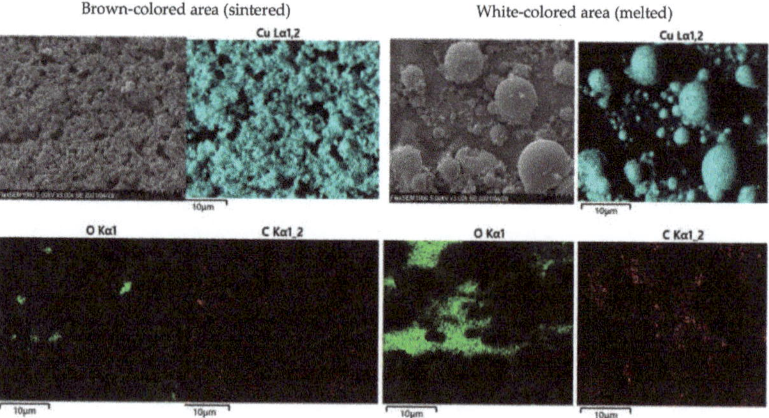

Figure 6. SEM images and EDS mappings of patterns fabricated under Ar gas injection rates of 125 mL/min.

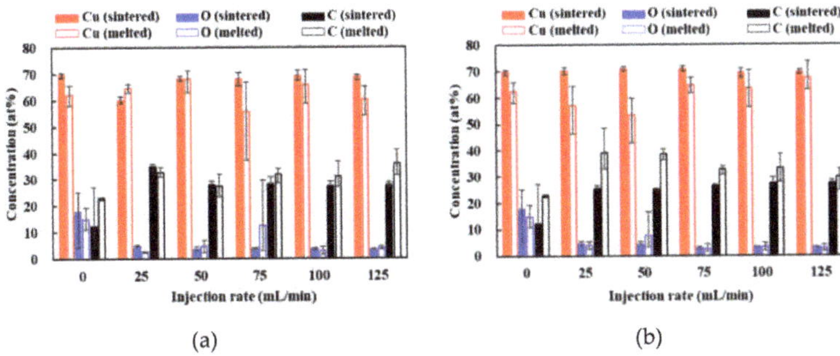

Figure 7. Concentration ratio of the patterns in sintered and melted area under (**a**) N_2 and (**b**) Ar gas injections.

3.5. Resistance

The resistances of the patterns were determined along their cross-sections. Figure 8a shows the height of the patterns fabricated at various injection rate of Ar gas. The resistance of the patterns fabricated under various gas injection rates is shown in Figure 8b. The resistance significantly increased over 100 mL/min of the Ar gas. This result qualitatively agreed with the ratio of sintered/melted area of the patterns as shown in Table 1. By improving the density by optimizing a concentration of CuO NPs or a size distribution of the CuO NPs in the CuO NP paste, the higher electrical conductivity of the patterns is expected to be formed by femtosecond laser reductive sintering of CuO NPs.

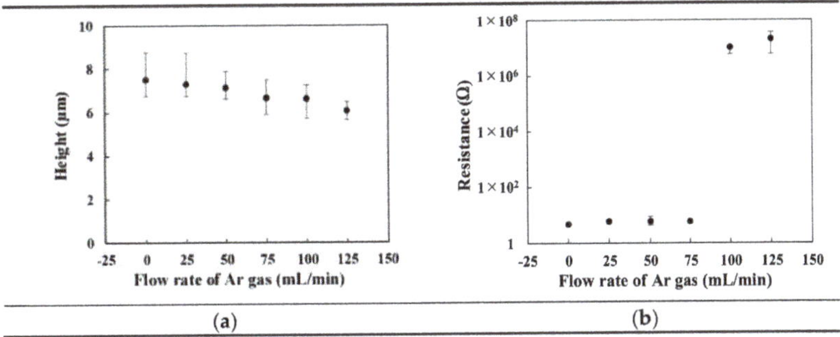

Figure 8. (**a**) Height and (**b**) resistance of the patterns fabricated under various injection rate of Ar gas.

4. Discussion

The concentration of oxygen decreased with increasing the gas injection rate by XRD and EDS analyses. In contrast, the resistance of the patterns fabricated under high gas injection rate of ≥ 100 mL/min, increased with increasing the gas injection rate. By considering the porosity of the patterns, the high resistance of the patterns was induced by the low density of the patterns. These results indicate that the purity of the patterns was improved by increasing the gas injection rate.

The advantage of the femtosecond laser reductive sintering is that the CuO NPs were reductive-sintered under low pulse energy of ~0.5 nJ, namely low average power of ~50 mW, by comparing to the use of continuous-wave and nanosecond pulsed lasers of ~0.2 W [14]. However, the concentration of the oxide in the patterns fabricated using femtosecond laser pulses [15,17,21] was higher than the patterns fabricated using continuous-wave and nanosecond pulsed lasers by examining the fabricated patterns [14]. We experimentally demonstrated that the high concentration of oxygen in the patterns was caused by the atmospheric oxygen by controlling the ambient atmosphere in the femtosecond laser irradiation process. This result is believed to be important to obtain Cu patterns with high purity and to consider the reduction process of CuO nanoparticles. In the femtosecond laser irradiation process, a large intensity of the femtosecond laser pulses is advantageous to heat the CuO nanoparticles rapidly. In contrast, the rapid heating and cooling processes generated the non-equilibrium material of Cu_2O in air [15,17,21]. The control of the ambient atmosphere is effective to improve the purity of the Cu in the patterns.

5. Conclusions

We evaluated the effect of the inert gas injections on the patterns fabricated by femtosecond laser reductive sintering of CuO nanoparticles.

(1) The XRD and EDS analysis suggest that the purity of the Cu in the patterns was improved by increasing the gas injection rate.

(2) The resistance of the patterns fabricated under excessively high injection rates ($\geq$100 mL/min) increased. By considering the density of the patterns, the high resistance was caused by the low density of the patterns and not high oxidation.

These results suggest that the rapid heating and cooling in the femtosecond laser reductive sintering of CuO NPs generated the non-equilibrium materials of Cu_2O due to the reoxidation in air, which was different from the irradiation of continuous-wave and nanosecond pulsed lasers to the CuO NP inks. The concentration of the oxygen in the patterns can be reduced by irradiating the femtosecond laser pulses under inert gas injection.

Author Contributions: Conceptualization, M.M.; methodology, M.M. and K.Y.; validation, M.M. and K.Y.; formal analysis, M.M. and K.Y.; investigation, K.Y.; resources, M.M.; data curation, M.M. and K.Y.; writing—original draft preparation, M.M. and K.Y.; writing—review and editing, M.M.; visualization, M.M. and K.Y.; supervision, M.M.; project administration, M.M.; funding acquisition, M.M. Both authors have read and agreed to the published version of the manuscript.

Funding: This work was supported by Funds for the Development of Human Resources in Science and Technology "2019 Initiative for Realizing Diversity in the Research Environment (Leading Type)"; The Ministry of Education, Culture, Sports, Science, and Technology (MEXT); and KAKENHI 20H02043.

Institutional Review Board Statement: Not applicable.

Informed Consent Statement: Not applicable.

Conflicts of Interest: The authors declare no conflict of interest.

References

1. Zacharatos, F.; Theodorakos, I.; Karvounis, P.; Tuohy, S.; Braz, N.; Melamed, S.; Kabla, A.; De la Vega, F.; Andritsos, K.; Hatziapostolou, A.; et al. Selective Laser Sintering of Laser Printed Ag Nanoparticle Micropatterns at High Repetition Rates. *Materials* **2018**, *11*, 2142. [CrossRef]
2. Yang, L.; Gan, Y.; Zhang, Y.; Chen, J.K. Molecular dynamics simulation of neck growth in laser sintering of different-sized gold nanoparticles under different heating rates. *Appl. Phys. A* **2011**, *106*, 725–735. [CrossRef]
3. Back, S.; Kang, B. Low-cost optical fabrication of flexible copper electrode via laser-induced reductive sintering and adhesive transfer. *Opt. Lasers Eng.* **2018**, *101*, 78–84. [CrossRef]
4. Halonen, E.; Heinonen, E.; Mäntysalo, M. The Effect of Laser Sintering Process Parameters on Cu Nanoparticle Ink in Room Conditions. *Opt. Photonics J.* **2013**, *3*, 40–44. [CrossRef]
5. Zenou, M.; Ermak, O.; Saar, A.; Kotler, Z. Laser sintering of copper nanoparticles. *J. Phys. D Appl. Phys.* **2014**, *47*, 025501. [CrossRef]
6. Cheng, C.W.; Chen, J.K. Femtosecond laser sintering of copper nanoparticles. *Appl. Phys. A* **2016**, *122*, 289. [CrossRef]
7. Soltani, A.; Khorramdel Vahed, B.; Mardoukhi, A.; Mantysalo, M. Laser sintering of copper nanoparticles on top of silicon substrates. *Nanotechnology* **2016**, *27*, 035203. [CrossRef]
8. Ohishi, T.; Kimura, R. Fabrication of Copper Wire Using Glyoxylic Acid Copper Complex and Laser Irradiation in Air. *Mater. Sci. Appl.* **2015**, *6*, 799–808. [CrossRef]

9. Ohishi, T.; Takahashi, N. Preparation and Properties of Copper Fine Wire on Polyimide Film in Air by Laser Irradiation and Mixed-Copper-Complex Solution Containing Glyoxylic Acid Copper Complex and Methylamine Copper Complex. *Mater. Sci. Appl.* **2018**, *9*, 859–872. [CrossRef]
10. Yu, J.H.; Rho, Y.; Kang, H.; Jung, H.S.; Kang, K.-T. Electrical behavior of laser-sintered Cu based metal-organic decomposition ink in air environment and application as current collectors in supercapacitor. *Int. J. Precis. Eng. Manuf.-Green Technol.* **2015**, *2*, 333–337. [CrossRef]
11. Mizoshiri, M.; Kondo, Y. Direct writing of Cu-based fine micropatterns using femtosecond laser pulse-induced sintering of Cu_2O nanospheres. *Jpn. J. Appl. Phys.* **2019**, *58*, SDDF05. [CrossRef]
12. Mizoshiri, M.; Kondo, Y. Direct writing of two- and three-dimensional Cu-based microstructures by femtosecond laser reductive sintering of the Cu_2O nanospheres. *Opt. Mater. Express* **2019**, *9*, 2828–2837. [CrossRef]
13. Lee, H.; Yang, M. Effect of solvent and PVP on electrode conductivity in laser-induced reduction process. *Appl. Phys. A* **2015**, *119*, 317–323. [CrossRef]
14. Kang, B.; Han, S.; Kim, J.; Ko, S.; Yang, M. One-Step Fabrication of Copper Electrode by Laser-Induced Direct Local Reduction and Agglomeration of Copper Oxide Nanoparticle. *J. Phys. Chem. C* **2011**, *115*, 23664–23670. [CrossRef]
15. Mizoshiri, M.; Ito, Y.; Arakane, S.; Sakurai, J.; Hata, S. Direct fabrication of Cu/Cu_2O composite micro-temperature sensor using femtosecond laser reduction patterning. *Jpn. J. Appl. Phys.* **2016**, *55*, 06GP05. [CrossRef]
16. Arakane, S.; Mizoshiri, M.; Hata, S. Direct patterning of Cu microstructures using femtosecond laser-induced CuO nanoparticle reduction. *Jpn. J. Appl. Phys.* **2015**, *54*, 06FP07. [CrossRef]
17. Mizoshiri, M.; Arakane, S.; Sakurai, J.; Hata, S. Direct writing of Cu-based micro -temperature detectors using femtosecond laser reduction of CuO nanoparticles. *Appl. Phys. Express* **2016**, *9*, 036701. [CrossRef]
18. Arakane, S.; Mizoshiri, M.; Sakurai, J.; Hata, S. Direct writing of three-dimensional Cu-based thermal flow sensors using femtosecond laser-induced reduction of CuO nanoparticles. *J. Micromech. Microeng.* **2017**, *27*, 055013. [CrossRef]
19. Roth, G.-L.; Haubner, J.; Kefer, S.; Esen, C.; Hellmann, R. Fs-laser based hybrid micromachining for polymer micro-opto electrical systems. *Opt. Lasers Eng.* **2021**, *137*, 106362. [CrossRef]
20. Rahman, M.K.; Lu, Z.; Kwon, K.-S. Green laser sintering of copper oxide (CuO) nano particle (NP) film to form Cu conductive lines. *AIP Adv.* **2018**, *8*, 095008. [CrossRef]
21. Huang, Y.; Xie, X.; Li, M.; Xu, M.; Long, J. Copper circuits fabricated on flexible polymer substrates by a high repetition rate femtosecond laser-induced selective local reduction of copper oxide nanoparticles. *Opt. Express* **2021**, *29*, 4453–4463. [CrossRef] [PubMed]
22. Mizoshiri, M.; Iijima, Y.; Hata, S. Preparation of Nonspherical Monodisperse Polydimethylsiloxane Microparticles for Self-assembly Fabrication of Periodic Structures. *IEEJ Trans. Sens. Micromach.* **2019**, *139*, 132–136. [CrossRef]

 *materials*

Article

Copper and Nickel Microsensors Produced by Selective Laser Reductive Sintering for Non-Enzymatic Glucose Detection

Ilya I. Tumkin [1,*], Evgeniia M. Khairullina [1], Maxim S. Panov [1], Kyohei Yoshidomi [2] and Mizue Mizoshiri [2]

[1]   Institute of Chemistry, Saint Petersburg State University, 7/9 Universitetskaya Nab.,
      199034 St. Petersburg, Russia; e.khayrullina@spbu.ru (E.M.K.); m.s.panov@spbu.ru (M.S.P.)
[2]   Department of Mechanical Engineering, Nagaoka University of Technology, Nagaoka 9402142, Japan;
      s193097@stn.nagaokaut.ac.jp (K.Y.); mizoshiri@mech.nagaokaut.ac.jp (M.M.)
*   Correspondence: i.i.tumkin@spbu.ru

Abstract: In this work, the method of selective laser reductive sintering was used to fabricate the sensor-active copper and nickel microstructures on the surface of glass-ceramics suitable for non-enzymatic detection of glucose. The calculated sensitivies for these microsensors are 1110 and 2080 μA mM^{-1}·cm^{-2} for copper and nickel, respectively. Linear regime of enzymeless glucose sensing is provided between 0.003 and 3 mM for copper and between 0.01 and 3 mM for nickel. Limits of glucose detection for these manufactured micropatterns are equal to 0.91 and 2.1 μM for copper and nickel, respectively. In addition, the fabricated materials demonstrate rather good selectivity, long-term stability and reproducibility.

Keywords: selective laser sintering; femtosecond laser; copper; nickel; microsensors; non-enzymatic sensing; glucose

Citation: Tumkin, I.I.; Khairullina, E.M.; Panov, M.S.; Yoshidomi, K.; Mizoshiri, M. Copper and Nickel Microsensors Produced by Selective Laser Reductive Sintering for Non-Enzymatic Glucose Detection. *Materials* **2021**, *14*, 2493. https://doi.org/10.3390/ma14102493

Academic Editor: Daniela Iannazzo

Received: 15 April 2021
Accepted: 7 May 2021
Published: 12 May 2021

Publisher's Note: MDPI stays neutral with regard to jurisdictional claims in published maps and institutional affiliations.

1. Introduction

At present, significant progress in the treatment of many human diseases including diabetes and cancer has been made. Nevertheless, the problem of timely diagnostics of a disease at the initial stages [1,2] and further adequate treatment have become more important. In this regard, particular attention attracts selective identification of various bioanalytes (chemical biomarkers) in human biological fluids and tissues; biosensors are the most effective devices applied for this purpose. Among the many different types of biosensors, electrochemical biosensors are the most distinguished exhibiting low cost, rapid analysis and simple sample preparation [3–5]. These devices are useful in sensing of such analytes as folic acid [6], DNA [7], D- and L-amino acids [8], uric acid [9], choline [10], many cancer biomarkers, e.g., cancer-embryonic antigen [11], glucose [12], hydrogen peroxide [13] and others. A large number of works deal with design and development of electrochemical and optical sensors, in which electrocatalytic activity is revealed by the enzyme-free compounds and materials. Indeed, the enzyme-free sensors allows for avoiding the problems in biosensorics caused by the utilization of enzymes (instability in aggressive environments, high cost and the need to immobilize enzymes on some carriers) [14]. The variety of materials can be applied for non-enzymatic sensing [15–23]. In this paper, we discuss sensors based on metals and their alloys and/or composites that have deserved a great interest in this scientific field.

It is necessary to mention the methods that are typically used for fabrication of such materials. In recent, 3D printing technologies have become a popular tool for solving a number of scientific problems [24]. Selective laser sintering (SLS) and selective laser manufacturing (SLM) for 3D microstructures are very promising representatives of such modern techniques [25]. In general, SLS is an additive manufacturing method that can be implemented for formation of 3D structures of a given size and shape by heating powders of various materials (plastic, glass, ceramics, metals) with the focused laser beam up to temperature, at which the powder-like particles fuse together providing the porosity control regime. In contrast to SLS, in SLM a powder is heated before the melting point that allows for decreasing the porosity and obtaining a homogeneous system. Therefore, as a rule, SLM is commonly used for fabrication of metallic materials. In this case, this method is recognized as direct laser metal sintering (DLMS). SLM is already actively applied in production of the electrochemical sensors based on metals and their alloys suitable for detection of phenols [26] or explosives and nerve agents [27]. Other noticeable techniques that can be utilized for the aforementioned purposes are: roll-to-roll printing (R2R) [28], inkjet printing [29], screen printing [30], chemical vapor deposition (CVD) [31], laser-induced metal deposition (LCLD) [32], femtosecond laser reductive sintering of metal oxide nanoparticles [25,33–35], etc. At some level all these methods demonstrate different drawbacks mostly related to cost, maintenance, complexity, efficiency and others. In turn, selective laser sintering, despite its own disadvantages, provides rather good reproducibility, rapidity and efficiency in fabrication of metallic structures reliable for non-enzymatic sensing. Femtosecond laser reductive sintering is a promising candidate for fabricating the metal patterns because the thick metal patterns can be formed in air by simulatenously metalizing and sintering of metal oxide nanoparticles. In comparison with LCLD, thicker metal patterns are formed by femtosecond laser pulse-induced thermochemical reaction because the intense laser pulses achieve the local heating of the raw metal oxide nanoparticles.

Thus, this work is devoted to the conditions optimization for fabrication of copper and nickel micropatterns using SLS method. We expect that the proposed approach can be applied as a fast, efficient and cheap way to create sensor platforms for enzyme-less detection of low concentrations of various bioanalytes. As a result, we fabricated the glucose microsensors based on such metals as copper (Cu) and nickel (Ni), which are widely used due to their low cost and high electrical conductivity.

2. Materials and Methods

The solution with CuO nanoparticles (NPs) (<50 nm particle size, Sigma Aldrich, St. Louis, MS, USA), polyvinylpyrrolidone (PVP, Mw 10000, Sigma Aldrich, St. Louis, MS, USA) and ethylene glycol (EG, 99.8%, Sigma Aldrich, St. Louis, MS, USA). First, PVP were mixed with EG. Subsequently, CuO NPs were dispersed into the mixed solution. The resulting CuO NP-based solution was cover by spin-coater on glass-ceramics substrates (5 s—500 rpm, 30 s—7000 rpm). A similar approach was applied to obtain films based on nickel oxide (NiO) nanoparticles. NiO NP (<50 nm particle size, Sigma Aldrich) were used as a source of Ni. It has been demonstrated that NiO NPs were reduced to Ni by femtosecond laser reduction of NiO NP solutions consisting of NiO NPs, PVP and EG 23.

All metal structures were produced on glass-ceramic. Glass-ceramic material consists of silicon dioxide (60.5%), aluminum oxide (13.5%), calcium oxide (8.5%), magnesium oxide (7.5%) and titanium dioxide (10.0%).

Surface preparation was carried out using ultrasonic cleaning in acetone, ethanol and water (sequentially), then the dried substrates were treated by irradiating ozone for 1 min to improve the wetting property of the prepared colloidal solution with polarity for each substrate using the FLAT EXCIMER EX-mini (Hamamatsu, Japan), after which the solution was deposited using a spin-coater (MS-B100, MIKASA CO. LTD, Tokyo, Japan).

A femtosecond fiber laser (TOPTICA Photonics AG, Munich, Germany), pulse duration: 120 fs, wavelength: 780 nm, repetition rate: 80 MHz) was used for direct patterning of Cu and Ni microstructures. Femtosecond laser pulses were focused with an objective lens with a numerical aperture of 0.45. First, a solution containing nanoparticles of CuO or NiO was placed on glass ceramics using spin-coating technique. Then, the micropatterns were directly produced by the focused femtosecond laser pulses. The raster pitch of the micropatterns was decided to be 5 μm and 10 μm.

The crystal structures of the patterns formed by raster scanning of the focused femtosecond laser pulses were examined by an X-ray diffraction (XRD) analysis (Rigaku RINT, MiniFlex) (Rigaku Corporation, Tokyo, Japan) using Cu Kα radiation.

The topology of Cu and Ni electrodes were investigated using scanning electron microscopy (Zeiss Supra 40 VP, Oberkochen, Germany). The EDX-system was coupled with a scanning electron microscope equipped with X-ray attachment (Oxford Instruments INCA X-act) (Oxford Instruments, Abingdon, UK) was conducted to quantitatively investigate the major chemical composition ratio in the Cu and Ni electrodes.

The electrocatalytic activity of the fabricated Cu and Ni-based materials towards glucose was studied using voltammetric methods (potentiostat, Elins P30I) (Electrochemical Instruments Ltd., Chernogolovka, Russia). In order to increase the adhesion and lifetime of the synthesized materials, we drop-casted 10 μL of Nafion suspension (0.05 wt%) on the Cu- and Ni-based electrodes. All measurements were carried out at room temperature in a typical three-electrode system using a Pt wire as a counter electrode, a Ag/AgCl reference electrode, and pre-air-dried Cu and Ni-based microelectrodes as working electrodes. The solutions of D-glucose of different concentrations were added to background solution (0.1 M sodium hydroxide) with simultaneous stirring.

3. Results

The optimized compositions for deposition of Cu and Ni micropatterns on glass-ceramic surfaces are presented in Table 1. The regimes of the composition optimization for deposition were previously published for Cu patterns on the glass surface [34]. In the current work, we proposed and optimized a technique for Cu and Ni manufacturing on the surface of glass-ceramics. The main concept of the SLS experiment used in the current study is shown in Figure 1.

Table 1. The optimized compositions for deposition of Cu and Ni micropatterns on glass-ceramic.

Material of Electrode	CuO or NiO, g	PVP, g	EG, g
Cu	3	0.65	1.35
Ni	1.5	0.65	1.35

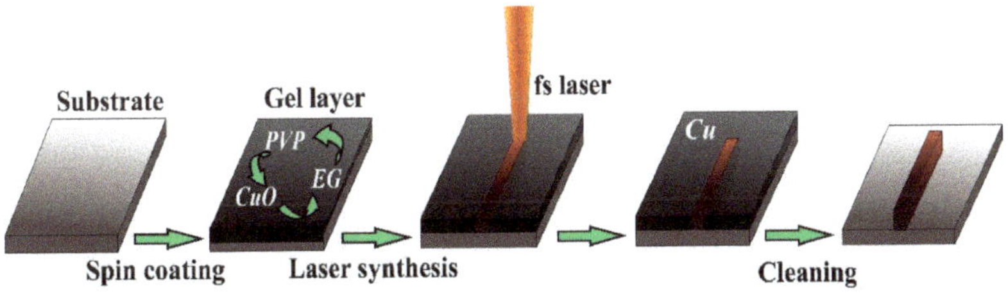

Figure 1. Schematic process of fabrication of metal micropatterns using SLS. A solution containing copper or nickel nanoparticles deposited on a substrate is irradiated with a femtosecond fiber laser (120 fs, Toptica, FemtoFiber pro NIR (780 nm)). The unreacted solution is removed by washing with a solvent.

In our previous works, it was shown that the main parameters that significantly affect the composition and topology of materials fabricated by the reductive SLS are laser fluence, scanning speed and pitch size. Therefore, by varying these parameters, it is possible to create materials with different functional properties. In this case, we have shown the possibility to produce the working electrodes for enzyme-free electrochemical detection of glucose. For this purpose, we optimized the conditions for synthesis of the conductive coatings on the surface of a dielectric with high adhesion and a developed surface. For fabrication of Cu electrodes on glass, we used the laser fluence varied from 0.0096 to 0.0230 J/cm^2, whereas the scanning speed was 1–10 mm per second and the distance between the deposited lines was varied between 5 and 10 μm. In turn, the analysis of SEM images of Cu patterns on glass-ceramics has shown that the most optimal conditions for producing homogeneous materials are the following: laser fluence of 0.0192 J/cm^2, scanning speed of 5 mm per second and distance between lines of 5 μm (Figure 2a). Thus, we were able to produce the metallic film consisting of the 50–300 nm Cu particles. It is important that the most homogeneous film on the surface of glass-ceramics can be deposited at a laser fluence of 0.0192 J/cm^2 (Figure 2); presumably, due to the more resistant to temperature nature of glass-ceramics in opposite to glass, for which the same value was equal to 0.0154 J/cm^2. Analysis of the electronic microphotographs of Ni patterns on the surface of glass-ceramics showed that the most homogeneous layer of metal could be obtained at laser fluence of 0.0192 J/cm^2 (Figure 3). In this regard, we have conducted the SLS experiments at laser fluence varied between 0.0096 and 0.0230 J/cm^2. At 0.0192 J/cm^2, Ni is deposited as a thin film consisting of particles with size of 100–500 nm. EDX analysis confirms that the main component of the patterns is Ni.

Figure 2. SEM images and results of EDX analysis (right top corner in each panel) of Cu patterns fabricated on glass-ceramics using SLS at the laser fluence of 0.0192 J/cm^2 along with the scanning speed of 5 (**a**) and 10 (**b**) mm per second.

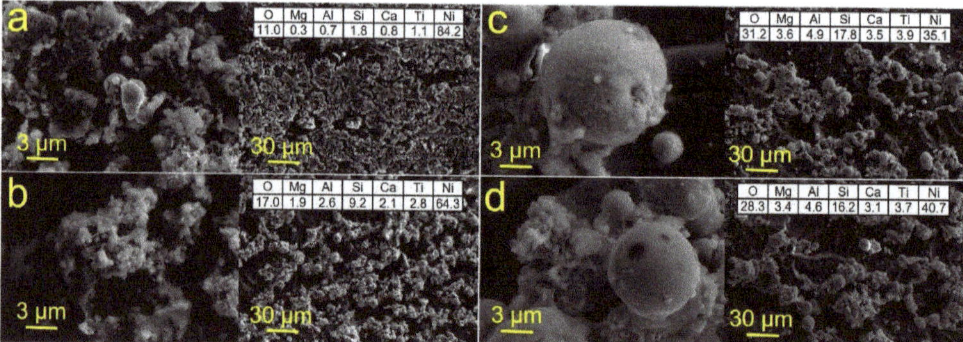

Figure 3. SEM images and results of EDX analysis (right top corner in each panel) of Ni patterns fabricated on glass-ceramics using SLS at the following laser fluences (J/cm^2) and scanning speeds (mm s^{-1}): (**a**) 0.0192 and 5; (**b**) 0.0154 and 5; (**c**) 0.0192 and 10; (**d**) 0.0154 and 10.

Summarizing the results of SEM and EDX studies, we found the most optimal conditions for Cu and Ni sintering (Table 2).

Table 2. The optimized SLS experimental conditions used for synthesis of Cu and Ni on glass-ceramics surfaces.

Material of Electrode	Laser Fluence, J/cm^2	Scanning Speed, mm/s	Pitch Size, μm
Cu (Glass) [35]	0.0154	5	5
Cu (Sitall)	0.0192	5	5
Ni (Sitall)	0.0192	5	5

Furthermore, we carried out the phase analysis of Cu and Ni materials obtained using the conditions mentioned before (Figure 4). XRD shows that the synthesized deposits contain metallic Cu and Ni together with a small amount of their oxides. Besides, we also observe peaks associated with the material of a substrate (glass-ceramics). Figure 3a illustrates XRD diffractograms of Cu patterns. As is shown here, for fabrication of patterns enriched with Cu, we need to apply the following conditions: laser fluence of 0.0192 J/cm^2, scanning speed of 5 mm per second and distance between the lines of 5 μm (Figure 4a, 1). XRD of the sample obtained at a scanning speed of 10 mm per second (Figure 4a, 2) reveals the presence of Cu(I) oxide indicating that the reduction reaction of Cu(II) to its metallic state was incomplete. According to phase studies shown in Figure 4b, we can conclude that the most optimal conditions for fabrication of Ni-rich homogeneous patterns are the following: laser fluence of 0.0192 J/cm^2 and scanning speed of 5 mm per second.

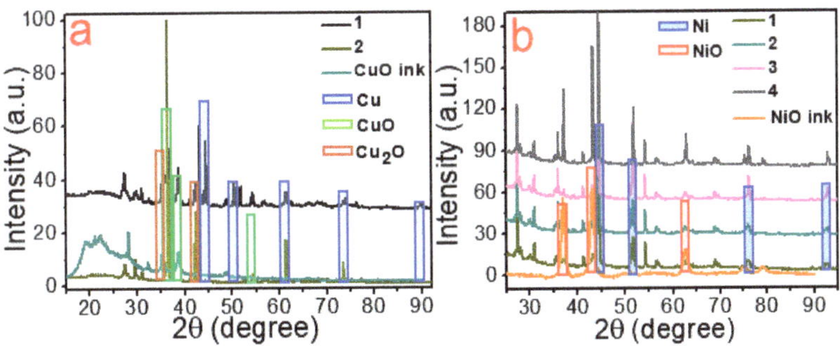

Figure 4. XRD patterns of (**a**) Cu (samples 1, 2 and initial mixture (CuO ink) are shown in the legend) on glass-ceramics and (**b**) Ni (samples 1–4 and initial mixture (NiO ink) are shown in the legend) on glass-ceramics sintered at the following laser fluence (J/cm^2), scanning speed (mm s^{-1}) and pitch size (μm): 0.0192, 5, 5 for 1; 0.0154, 5, 5 for 2; 0.0192, 10, 5 for 3; 0.0154, 10, 5 for 4.

It should be pointed out here that increasing the pitch size and the scanning speed (up to 10 mm per second) during laser irradiation leads to formation of non-uniform deposits with numerous defects containing metal oxides and, as a result, having a poor electrical conductivity. This could be related to the lack of time and laser fluence to complete metal reduction reaction (for both Cu and Ni). On the other hand, a too-low scanning speed (less than 1 mm per second) contributes to formation of smoother films due to a more complete fusion of the original NPs, resulting in deformation of a deposit and decrease of the surface area and the number of available active electrocatalytic centers.

Thus, using SLS, it is possible to fabricate the electrodes of various shapes and geometries relying on the optimized parameters. Figure 5 illustrates different geometries of copper microelectrodes obtained on the surface of glass-ceramics. Potentially, these geometries can be used to create a sensor platform based on the three-electrode electrochemical cell (Figure 5a).

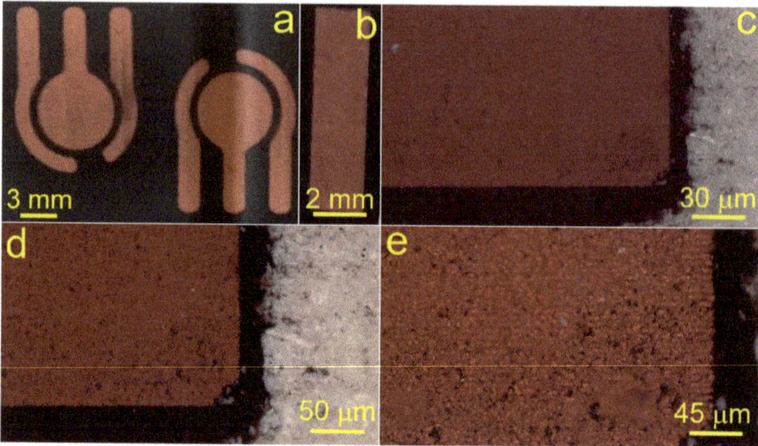

Figure 5. Photographs of copper micropatterns of different geometry fabricated on the surface of glass-ceramics using SLS. (**a**) Three-electrode geometry suitable for design of a sensor platform based on copper; (**b–e**) Pictures of copper rectangular microstructures of different scale. Black background corresponds to the area unexposed to the laser irradiation.

The electrocatalytic activity of the synthesized Cu and Ni micropatterns towards enzyme-free glucose sensing was tested using voltammetric methods. Figures 6a and 7a demonstrate cyclic voltammograms (CVs) of these materials measured in the background solution containing 1 mM D-glucose. The shape of CVs for Cu sintered on glass-ceramics have very broad range of potentials between 0.35 and 0.65 V corresponding to anodic glucose oxidation (Figure 6a). In turn, electrooxidation of glucose on Ni microelectrode takes place within the region of potentials of 0.45–0.7 V, which shifts toward larger potentials with an increase of the glucose concentration (Figure 7a).

All electrochemical characteristics such as sensitivity, limit of detection (LOD), linear range of glucose detection and selectivity were obtained using amperometry. Figures 6b and 7b illustrate the amperometric response to the consecutive additions of D-glucose to 0.1 M NaOH at potentials of 0.51 V for Cu and 0.6 V for Ni. Then, we obtained the linear dependence of the analytical signal (Faraday current) vs. D-glucose concentration for each material (Figures 6c and 7c). According to this data, linear regime of enzymeless glucose detection is provided between 0.003 and 3 mM for Cu, whereas for Ni linear range lies between 0.01 and 3 mM. The sensitivity of the electrodes were estimated by calculating the slopes of a linear curves shown in Figures 6c and 7c. As a result, the calculated sensitivities for Cu and Ni are 1110 and 2080 μA mM^{-1}·cm^{-2}, respectively. In addition, limits of glucose detection for all manufactured materials were calculated as LOD = 3 S/b, where S is the standard deviation from linearity and b is the slope of the calibration curve indicated in Figures 6c and 7c, and are equal to 0.91 and 2.1 μM for Cu and Ni, respectively. The measurement error does not exceed 7%. In addition, the error in the calculated sensitivities is also very small (R^2 is close to unity and is 0.9996 and 0.9994 for copper and nickel samples, respectively). Rather high sensitivity values can be explained by the close contact between the electrocatalytically active material and the current collector, since they are a single structure in contrast to the electrodes deposited on a conductive substrate by dropcasting or similar methods. This leads to an increase in electric conductivity and, as a result, sensitivity. The values of the electrochemical parameters observed for Cu and Ni micropatterns in this work in comparison with characteristics of the analogical non-enzymatic sensors are shown in Table 3.

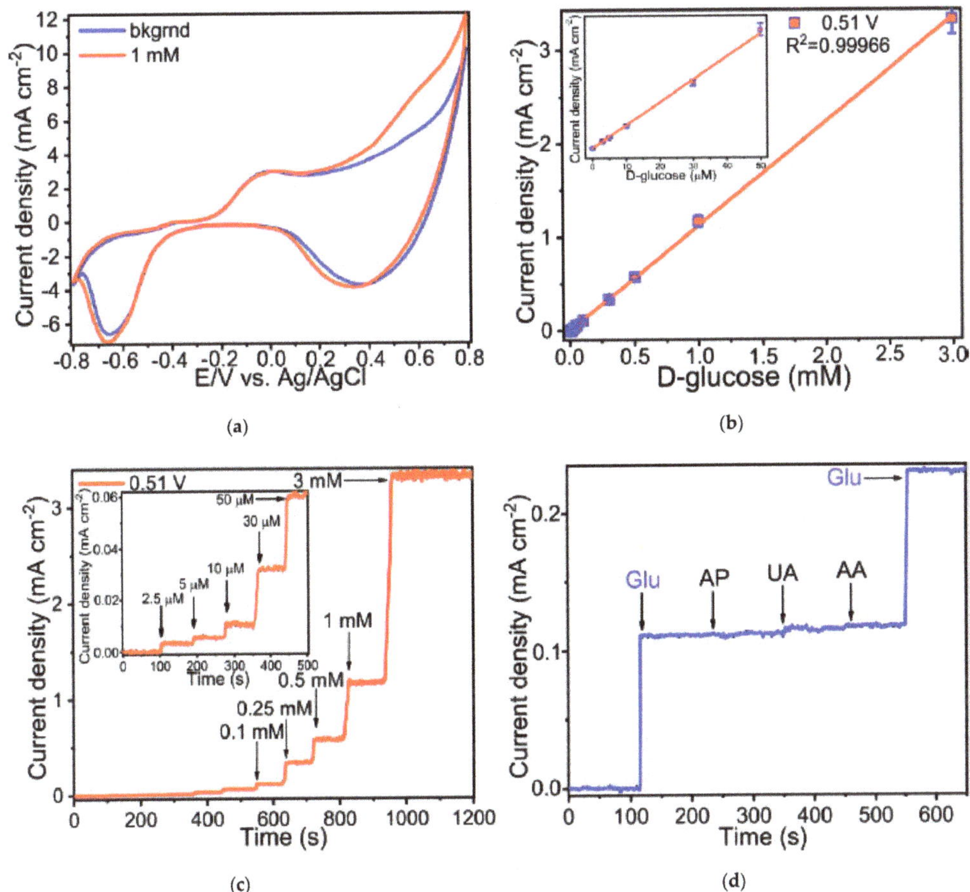

Figure 6. (**a**) Cyclic voltammograms of Cu patterns (electrode) measured in the solution of 0.1 M NaOH with addition of 1 mM D-glucose; (**b**) Amperogram of Cu electrode recorded at potential of 0.51 V in 0.1 M NaOH solution containing of D-glucose of various concentration; (**c**) Linear range of D-glucose concentrations for enzyme-less sensing using the fabricated Cu electrode; (**d**) Selectivity of Cu electrode towards 100 μM D-glucose (Glu) detection in the presence of 20 μM 4-acetamidophenol (AP), 20 μM uric acid (UA) and 20 μM ascorbic acid (AA) observed in the background solution of 0.1 M NaOH. In these experiments, we used Cu electrodes sintered on glass-ceramics.

The selectivity of the fabricated materials with respect to glucose sensing was investigated in the presence of such interfering compounds as 4-acetamidophenol (AP), ascorbic acid (AA) and uric acid (UA) that usually coexist with glucose in the human blood (Figures 6d and 7d). Thus, Cu and Ni micropatterns have a good selectivity for glucose sensing exhibiting much more significant analytical response towards D-glucose opposite to other analytes.

We also studied the long-term stability and reproducibility of the fabricated microelectrodes. Rather good stability was confirmed by testing five samples of each material for 10 days. We observed that during this period all samples maintained ~92–95% their initial electrocatalytic activity with respect to non-enzymatic glucose sensing. On the other hand, the great reproducibility was supported by low values of the relative standard deviation (~5–8% for all samples) of the analytical response to 1 mM D-glucose.

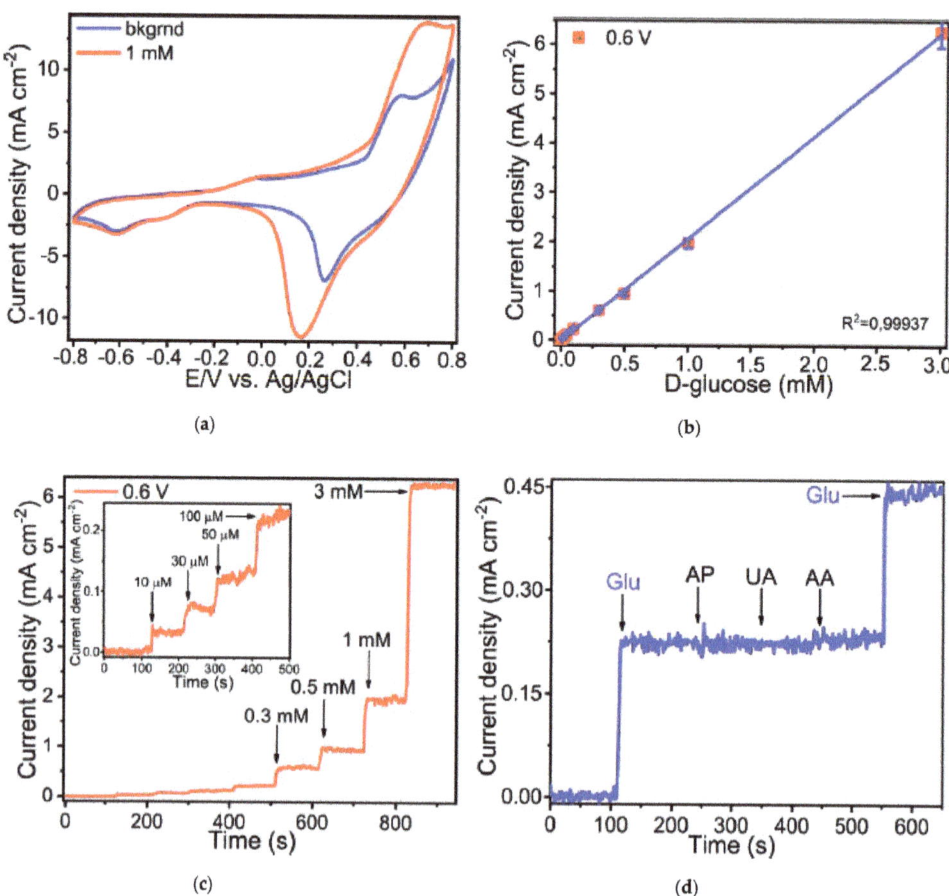

Figure 7. (**a**) Cyclic voltammograms of Ni micropatterns (electrode) measured in the solution of 0.1 M NaOH with addition of 1 mM D-glucose; (**b**) Amperogram of Ni electrode recorded at potential of 0.6 V in 0.1 M NaOH solution containing of D-glucose of various concentration; (**c**) Linear range of D-glucose concentrations for enzyme-less sensing using the fabricated Ni electrode; (**d**) Selectivity of Ni electrode towards 100 µM D-glucose (Glu) detection in the presence of 20 µM 4-acetamidophenol (AP), 20 µM uric acid (UA) and 20 µM ascorbic acid (AA) observed in the background solution of 0.1 M NaOH. In these experiments, we used Ni electrodes sintered on glass-ceramics.

Table 3. Different electrode materials for non-enzymatic glucose sensing in comparison with those fabricated in this work.

Electrode Material	Sensitivity (μA mM^{-1} cm^{-2})	Linear Range (mM)	Limit of Detection (μM)	Ref.
Cu on glass-ceramics	1110 ± 6,45	0.003−3	0.91	This work
Ni on glass-ceramics	2080 ± 18,53	0.01−3	2.1	This work
Cu MPs	2432	0−4.711	0.19	[36]
Cu coating	2149.1	0.001−4.6	0.03	[37]
carbon electrode/nanoporous Cu	33.75	0.0006−3.369	2.6	[38]
Cu NPs	412	0−0.7	2.76	[39]
Ni NPs on carbon nanotubes	1438	0.001−1	0.5	[40]
rhizobia-like Ni NPs	50.97	0.001−7	0.18	[41]
Ni NP/chitosan	318.4	0−9	4.1	[42]
3D porous carbon/Ni NPs	207.3	0.015−6.45	4.8	[43]

4. Conclusions

In this work, we found the optimal parameters for manufacturing Cu and Ni micropatterns on the surface of glass-ceramics using reductive selective laser sintering (SLS). In order to increase the adhesion and lifetime of these materials on the surfaces of the substrate used in the current study, they were treated with Nafion solution. The fabricated Cu and Ni electrode materials can be used for non-enzymatic glucose sensing. It was confirmed by the electrochemical experiments that revealed their high sensitivity (1110 and 2080 $\mu A\ mM^{-1} \cdot cm^{-2}$), low limit of detection (0.91 and 2.1 μM), broad linear range (0.003–3 mM and 0.01–3 mM), as well as good selectivity and long-run stability. Thus, it is possible to conclude that high speed reductive SLS allowing for obtaining electrodes of various geometries is a quite promising technique for the design and fabrication of reliable materials for enzyme-free sensing purposes that can compete with existing technologies used for the production of microelectronic devices and sensors.

Author Contributions: Conceptualization, Writing—original draft, funding acquisition, I.I.T.; Visualization, Conceptualization, Investigation, E.M.K.; Visualization, Writing—original draft, M.S.P.; Investigation, K.Y.; Conceptualization, Writing—review & editing, supervision, funding acquisition, M.M. All authors have read and agreed to the published version of the manuscript.

Funding: This research was funded by RFBR-JSPS Bilateral Joint Research Projects, RFBR project number 20-53-50011 and JSPS project number JSPSBP120204807 and the APC was funded by personal funding of researcher.

Data Availability Statement: The data presented in this study are available on request from the corresponding author.

Acknowledgments: The authors would like to thank the SPbSU Nanotechnology Interdisciplinary Centre, Centre for Optical and Laser Materials Research and Centre for X-ray Diffraction Studies.

Conflicts of Interest: The authors declare no conflict of interest.

References

1. Tørring, M.L.; Frydenberg, M.; Hamilton, W.; Hansen, R.P.; Lautrup, M.D.; Vedsted, P. Diagnostic interval and mortality in colorectal cancer: U-shaped association demonstrated for three different datasets. *J. Clin. Epidemiol.* **2012**, *65*, 669–678. [CrossRef]
2. Sun, J.Y.; Shi, Y.; Cai, X.Y.; Liu, J. Potential diagnostic and therapeutic value of circular RNAs in cardiovascular diseases. *Cell Signal.* **2020**, *71*, 109604. [CrossRef] [PubMed]
3. Cesewski, E.; Johnson, B.N. Electrochemical biosensors for pathogen detection. *Biosens. Bioelectron.* **2020**, *159*, 112214. [CrossRef]
4. Park, S.S.; Tatum, C.E.; Lee, Y. Dual electrochemical microsensor for simultaneous measurements of nitric oxide and oxygen: Fabrication and characterization. *Electrochem. Commun.* **2009**, *11*, 2040–2043. [CrossRef]
5. Nasiri, S.; Khosravani, M.R. Progress and challenges in fabrication of wearable sensors for health monitoring. *Sens. Actuators A Phys.* **2020**, *312*, 112105. [CrossRef]
6. Batra, B.; Narwal, V.; Kalra, V.; Sharma, M.; Rana, J.S. Folic acid biosensors: A review. *Process Biochem.* **2020**, *92*, 343–354. [CrossRef]
7. Blair, E.O.; Corrigan, D.K. A review of microfabricated electrochemical biosensors for DNA detection. *Biosens. Bioelectron.* **2019**, *134*, 57–67. [CrossRef]
8. Pundir, C.S.; Lata, S.; Narwal, V. Biosensors for determination of D and L- amino acids: A review. *Biosens. Bioelectron.* **2018**, *117*, 373–384. [CrossRef] [PubMed]
9. Pundir, C.S.; Jakhar, S.; Narwal, V. Determination of urea with special emphasis on biosensors: A review. *Biosens. Bioelectron.* **2019**, *123*, 36–50. [CrossRef] [PubMed]
10. Rahimi, P.; Joseph, Y. Enzyme-based biosensors for choline analysis: A review. *TrAC Trends Anal. Chem.* **2019**, *110*, 367–374. [CrossRef]
11. Khanmohammadi, A.; Aghaie, A.; Vahedi, E.; Qazvini, A.; Ghanei, M.; Afkhami, A.; Hajian, A.; Bagheri, H. Electrochemical biosensors for the detection of lung cancer biomarkers: A review. *Talanta* **2020**, *206*, 120251. [CrossRef] [PubMed]
12. Sehit, E.; Altintas, Z. Significance of nanomaterials in electrochemical glucose sensors: An updated review (2016-2020). *Biosens. Bioelectron.* **2020**, *159*, 112165. [CrossRef]
13. Liu, W.; Zhang, H.; Yang, B.; Li, Z.; Lei, L.; Zhang, X. A non-enzymatic hydrogen peroxide sensor based on vertical NiO nanosheets supported on the graphite sheet. *J. Electroanal. Chem.* **2015**, *749*, 62–67. [CrossRef]
14. Revathi, C.; Rajendra Kumar, R.T. *Enzymatic and Nonenzymatic Electrochemical Biosensors*; Elsevier Ltd.: Amsterdam, The Netherlands, 2019; ISBN 9780081025772.
15. Gong, X.; Gu, Y.; Zhang, F.; Liu, Z.; Li, Y.; Chen, G.; Wang, B. High-performance non-enzymatic glucose sensors based on CoNiCu alloy nanotubes arrays prepared by electrodeposition. *Front. Mater.* **2019**, *6*, 1–9. [CrossRef]

16. Wang, L.; Zhu, W.; Lu, W.; Qin, X.; Xu, X. Surface plasmon aided high sensitive non-enzymatic glucose sensor using Au/NiAu multilayered nanowire arrays. *Biosens. Bioelectron.* **2018**, *111*, 41–46. [CrossRef] [PubMed]

17. Liu, X.; Yang, W.; Chen, L.; Jia, J. Three-Dimensional Copper Foam Supported CuO Nanowire Arrays: An Efficient Non-enzymatic Glucose Sensor. *Electrochim. Acta* **2017**, *235*, 519–526. [CrossRef]

18. S. Panov, M.; M. Khairullina, E.; S. Vshivtcev, F.; N. Ryazantsev, M.; I. Tumkin, I. Laser-Induced Synthesis of Composite Materials Based on Iridium, Gold and Platinum for Non-Enzymatic Glucose Sensing. *Materials* **2020**, *13*, 3359. [CrossRef] [PubMed]

19. Panov, M.S.; Vereshchagina, O.A.; Ermakov, S.S.; Tumkin, I.I.; Khairullina, E.M.; Skripkin, M.Y.; Mereshchenko, A.S.; Ryazantsev, M.N.; Kochemirovsky, V.A. Non-enzymatic sensors based on in situ laser-induced synthesis of copper-gold and gold nano-sized microstructures. *Talanta* **2017**, *167*, 201–207. [CrossRef] [PubMed]

20. Panov, M.; Aliabev, I.; Khairullina, E.; Mironov, V.; Tumkin, I. Fabrication of nickel-gold microsensor using in situ laser-induced metal deposition technique. *J. Laser Micro Nanoeng.* **2019**, *14*, 266–269. [CrossRef]

21. Smikhovskaia, A.V.; Andrianov, V.S.; Khairullina, E.M.; Lebedev, D.V.; Ryazantsev, M.N.; Panov, M.S.; Tumkin, I.I. In situ laser-induced synthesis of copper-silver microcomposite for enzyme-free D-glucose and L-alanine sensing. *Appl. Surf. Sci.* **2019**, *488*, 531–536. [CrossRef]

22. Smikhovskaia, A.V.; Panov, M.S.; Tumkin, I.I.; Khairullina, E.M.; Ermakov, S.S.; Balova, I.A.; Ryazantsev, M.N.; Kochemirovsky, V.A. In situ laser-induced codeposition of copper and different metals for fabrication of microcomposite sensor-active materials. *Anal. Chim. Acta* **2018**, *1044*, 138–146. [CrossRef]

23. Baranauskaite, V.E.; Novomlinskii, M.O.; Tumkin, I.I.; Khairullina, E.M.; Mereshchenko, A.S.; Balova, I.A.; Panov, M.S.; Kochemirovsky, V.A. In situ laser-induced synthesis of gas sensing microcomposites based on molybdenum and its oxides. *Compos. Part B Eng.* **2019**, *157*, 322–330. [CrossRef]

24. Manzanares Palenzuela, C.L.; Pumera, M. (Bio)Analytical chemistry enabled by 3D printing: Sensors and biosensors. *TrAC Trends Anal. Chem.* **2018**, *103*, 110–118. [CrossRef]

25. Tamura, K.; Mizoshiri, M.; Sakurai, J.; Hata, S. Ni-based composite microstructures fabricated by femtosecond laser reductive sintering of NiO/Cr mixed nanoparticles. *Jpn. J. Appl. Phys.* **2017**, *56*. [CrossRef]

26. Cheng, T.S.; Nasir, M.Z.M.; Ambrosi, A.; Pumera, M. 3D-printed metal electrodes for electrochemical detection of phenols. *Appl. Mater. Today* **2017**, *9*, 212–219. [CrossRef]

27. Tan, C.; Nasir, M.Z.M.; Ambrosi, A.; Pumera, M. 3D Printed Electrodes for Detection of Nitroaromatic Explosives and Nerve Agents. *Anal. Chem.* **2017**, *89*, 8995–9001. [CrossRef] [PubMed]

28. Bariya, M.; Shahpar, Z.; Park, H.; Sun, J.; Jung, Y.; Gao, W.; Nyein, H.Y.Y.; Liaw, T.S.; Tai, L.C.; Ngo, Q.P.; et al. Roll-to-Roll Gravure Printed Electrochemical Sensors for Wearable and Medical Devices. *ACS Nano* **2018**, *12*, 6978–6987. [CrossRef]

29. Su, C.H.; Kung, C.W.; Chang, T.H.; Lu, H.C.; Ho, K.C.; Liao, Y.C. Inkjet-printed porphyrinic metal-organic framework thin films for electrocatalysis. *J. Mater. Chem. A* **2016**, *4*, 11094–11102. [CrossRef]

30. Li, M.; Li, Y.T.; Li, D.W.; Long, Y.T. Recent developments and applications of screen-printed electrodes in environmental assays-A review. *Anal. Chim. Acta* **2012**, *734*, 31–44. [CrossRef]

31. Barreca, D.; Massignan, C.; Daolio, S.; Fabrizio, M.; Piccirillo, C.; Armelao, L.; Tondello, E. Composition and microstructure of cobalt oxide thin films obtained from a novel cobalt(II) precursor by chemical vapor deposition. *Chem. Mater.* **2001**, *13*, 588–593. [CrossRef]

32. Kochemirovsky, V.A.; Skripkin, M.Y.; Tveryanovich, Y.S.; Mereshchenko, A.S.; Gorbunov, A.O.; Panov, M.S.; Tumkin, I.I.; Safonov, S. V Laser-induced copper deposition from aqueous and aqueous–organic solutions: State of the art and prospects of research. *Russ. Chem. Rev.* **2015**, *84*, 1059–1075. [CrossRef]

33. Arakane, S.; Mizoshiri, M.; Sakurai, J.; Hata, S. Three-dimensional Cu microfabrication using femtosecond laser-induced reduction of CuO nanoparticles. *Appl. Phys. Express* **2017**, *10*. [CrossRef]

34. Mizoshiri, M.; Nishitani, K.; Hata, S. Effect of heat accumulation on femtosecond laser reductive sintering of mixed CuO/NiO nanoparticles. *Micromachines* **2018**, *9*, 264. [CrossRef]

35. Mizoshiri, M.; Arakane, S.; Sakurai, J.; Hata, S. Direct writing of Cu-based micro-temperature detectors using femtosecond laser reduction of CuO nanoparticles. *Appl. Phys. Express* **2016**, *9*. [CrossRef]

36. Guo, M.M.; Xia, Y.; Huang, W.; Li, Z. Electrochemical fabrication of stalactite-like copper micropillar arrays via surface rebuilding for ultrasensitive nonenzymatic sensing of glucose. *Electrochim. Acta* **2015**, *151*, 340–346. [CrossRef]

37. Hou, L.; Zhao, H.; Bi, S.; Xu, Y.; Lu, Y. Ultrasensitive and highly selective sandpaper-supported copper framework for non-enzymatic glucose sensor. *Electrochim. Acta* **2017**, *248*, 281–291. [CrossRef]

38. Mei, L.; Zhang, P.; Chen, J.; Chen, D.; Quan, Y.; Gu, N.; Zhang, G.; Cui, R. Non-enzymatic sensing of glucose and hydrogen peroxide using a glassy carbon electrode modified with a nanocomposite consisting of nanoporous copper, carbon black and nafion. *Microchim. Acta* **2016**, *183*, 1359–1365. [CrossRef]

39. Shi, L.; Zhu, X.; Liu, T.; Zhao, H.; Lan, M. Encapsulating Cu nanoparticles into metal-organic frameworks for nonenzymatic glucose sensing. *Sens. Actuators B Chem.* **2016**, *227*, 583–590. [CrossRef]

40. Nie, H.; Yao, Z.; Zhou, X.; Yang, Z.; Huang, S. Nonenzymatic electrochemical detection of glucose using well-distributed nickel nanoparticles on straight multi-walled carbon nanotubes. *Biosens. Bioelectron.* **2011**, *30*, 28–34. [CrossRef]

41. Huo, K.; Fu, J.; Zhang, X.; Xu, P.; Gao, B.; Mooni, S.; Li, Y.; Fu, J. Phase separation induced rhizobia-like Ni nanoparticles and TiO2 nanowires composite arrays for enzyme-free glucose sensor. *Sens. Actuators B Chem.* **2017**, *244*, 38–46. [CrossRef]

42. Yang, J.; Yu, J.H.; Rudi Strickler, J.; Chang, W.J.; Gunasekaran, S. Nickel nanoparticle-chitosan-reduced graphene oxide-modified screen-printed electrodes for enzyme-free glucose sensing in portable microfluidic devices. *Biosens. Bioelectron.* **2013**, *47*, 530–538. [CrossRef] [PubMed]
43. Wang, L.; Zhang, Y.; Yu, J.; He, J.; Yang, H.; Ye, Y.; Song, Y. A green and simple strategy to prepare graphene foam-like three-dimensional porous carbon/Ni nanoparticles for glucose sensing. *Sens. Actuators B Chem.* **2017**, *239*, 172–179. [CrossRef]

Article

Laser-Induced Deposition of Plasmonic Ag and Pt Nanoparticles, and Periodic Arrays

Daria V. Mamonova [1], Anna A. Vasileva [1], Yuri V. Petrov [2], Denis V. Danilov [3], Ilya E. Kolesnikov [4], Alexey A. Kalinichev [4], Julien Bachmann [1,5] and Alina A. Manshina [1,*]

[1] Institute of Chemistry, Saint-Petersburg State University, 26 Universitetskii Prospect, Saint-Petersburg 198504, Russia; magwi@mail.ru (D.V.M.); anvsilv@gmail.com (A.A.V.); julien.bachmann@fau.de (J.B.)

[2] Department of Physics, Saint-Petersburg State University, Ulyanovskaya 3, Saint-Petersburg 198504, Russia; y.petrov@spbu.ru

[3] Interdisciplinary Resource Center for Nanotechnology, Research Park, Saint-Petersburg State University, Ulyanovskaya 1, Saint-Petersburg 198504, Russia; danilov1denis@gmail.com

[4] Centre for Optical and Laser Materials Research, Research Park, Saint-Petersburg State University, Ulyanovskaya 5, Saint-Petersburg 198504, Russia; ilya.kolesnikov@spbu.ru (I.E.K.); kalinichev.alex@gmail.com (A.A.K.)

[5] Department of Chemistry and Pharmacy, Friedrich–Alexander University of Erlangen—Nürnberg, IZNF, Cauerst. 3, 91058 Erlangen, Germany

* Correspondence: a.manshina@spbu.ru

Abstract: Surfaces functionalized with metal nanoparticles (NPs) are of great interest due to their wide potential applications in sensing, biomedicine, nanophotonics, etc. However, the precisely controllable decoration with plasmonic nanoparticles requires sophisticated techniques that are often multistep and complex. Here, we present a laser-induced deposition (LID) approach allowing for single-step surface decoration with NPs of controllable composition, morphology, and spatial distribution. The formation of Ag, Pt, and mixed Ag-Pt nanoparticles on a substrate surface was successfully demonstrated as a result of the LID process from commercially available precursors. The deposited nanoparticles were characterized with SEM, TEM, EDX, X-ray diffraction, and UV-VIS absorption spectroscopy, which confirmed the formation of crystalline nanoparticles of Pt (3–5 nm) and Ag (ca. 100 nm) with plasmonic properties. The advantageous features of the LID process allow us to demonstrate the spatially selective deposition of plasmonic NPs in a laser interference pattern, and thereby, the formation of periodic arrays of Ag NPs forming diffraction grating

Keywords: laser-induced deposition; noble metal NPs; plasmon resonance; nano-grating structures

Citation: Mamonova, D.V.; Vasileva, A.A.; Petrov, Y.V.; Danilov, D.V.; Kolesnikov, I.E.; Kalinichev, A.A.; Bachmann, J.; Manshina, A.A. Laser-Induced Deposition of Plasmonic Ag and Pt Nanoparticles, and Periodic Arrays. *Materials* **2021**, *14*, 10. http://dx.doi.org/10.3390/ma14010010

Received: 30 November 2020
Accepted: 21 December 2020
Published: 22 December 2020

1. Introduction

Laser fabrication methods provide specific physicochemical conditions in the zone of processing. As a result, they open efficient and highly controlled ways of nanomaterials synthesis with specific functionality. Metal nanoparticles (NPs) are a particular class of objects that are of interest due to their plasmonic properties and that can be synthesized with laser technologies. Among the most widespread techniques of metal NPs synthesis, are pulsed laser ablation/deposition in vacuum, gas, or liquid phases [1–3], as well as direct synthesis under intense laser beam focused in a solutions of metal salts [4,5]. These methods demonstrate high processing efficiency and a wide list of metal NPs have been synthesized with controlled parameters. However, their application for the surface functionalization with NPs has been difficult.

Other promising laser technologies rely on the creation of surfaces decorated/functionalized with metal NPs. One of them is the colorization of metal surfaces due to formation of plasmonic NPs by means of metal surface treatment with pulsed laser ablation [6]. Laser-induced forward transfer (LIFT) achieves the spatially controlled formation of a donor material (that can be a metal film as well) on a receiving substrate [7]. As a hybrid method of surface functionalization with metal NPs, preliminary surface activation with femtosecond laser beam, followed by laser induced photoreduction, can be mentioned [8].

All the listed methods are based on the use of highly intense pulsed laser radiation that imposes some restrictions to their applicability and the list of substrates that can be decorated with metal nanostructures. Among the methods combining both the ability of surface decoration with metal NPs and the use of a continuum-wave laser irradiation is the group of laser-induced depositions (LID). The LID approach is based on laser irradiation of the substrate–solution interface with a laser beam. As a result of laser-initiated chemical processes, metal structures are formed in the laser-affected area of the substrate. As a liquid phase, one can use electrolyte solutions traditional for chemical metallization [9–11], or solutions of metal salts with some reducing agents [12–16], or solutions of organometallic complexes [17,18]. The main difference between these approaches is the nature of the laser-induced process—the redox process initiated by laser heating (first case) or photochemical effects in the second and third variants. LID processes based on laser heating (often referred as laser-induced chemical liquid phase deposition of metals—LCLD) typically use laser irradiation of visible range with rather high power (hundreds of mW). Though LCLD allows precipitation of the functional, electrically conducting metal structures [19], its main limitation is the high-temperature process of reduction of metal ions and the side effects of local decomposition of precipitate and substrate. These side effects lead to inhomogeneous deposition of the metal phase, and as a result, rather high specific resistance of the deposits [20].

The photo-induced LID process uses laser irradiation of low intensity to photo-initiate the decomposition of an organometallic complex (that is the main component of the solution) dissolved in an appropriate solvent. As the next step, the formation of nanostructures from the precursor components takes place on the surface of the substrate. The main feature of the LID process is the spectral selectivity of the process (the laser wavelength should coincide with absorption band of the precursor), allowing one to use the low-intensity continuous-wave laser irradiation. The LID process proved to be very flexible, efficient, and well-controlled. It is characterized by good correlation of the chemical composition of the precursor and the deposited structures that is a consequence of the mild conditions of laser irradiation. The main competitive advantage of LID over other laser-based methods is its ability to synthesize NPs directly on any given/preselected surface area of the substrate [21,22]. It allows 'on-site' functionalization of various types of surfaces with NPs active in catalysis and electrocatalysis, surface-enhanced Raman spectroscopy (SERS), plasmon-enhanced fluorescence, energy conversion, solar cell technologies, etc. [17,18,23,24].

To date, the LID decoration of different surfaces has been successfully demonstrated—such as 2D substrates (microscope cover glasses and Si wafers both pristine and covered by indium tin oxide film) and 3D substrates (nanowires, capillaries, and porous anodic alumina membranes) [22,25,26].

As a main constraint of LID, one can consider the limited number of organometallic complexes demonstrated so far as LID precursor. In most cases, home-made supramolecular complexes were used that were heterometallic transition–metal polynuclear complexes. This kind of complex is of high interest for the LID process as they have precisely defined central bimetallic cluster core, allowing synthesis of bimetallic nanoparticles in the form of alloys. However, the preliminary synthesis procedure of home-made organometallic complexes can be considered as an obstacle to a widespread application of LID.

Here, we present the successful laser-induced deposition of monometallic Ag and Pt nanoparticles, and mixed Ag/Pt samples from commercially available precursors—organic salts, organometallic complexes without reducing agents adding to the solutions. This opens wide perspectives to LID and makes it generally available. We demonstrate precise control over the morphology and composition of the deposits by means of routine variation of the LID experimental parameters. As another LID advantage, we present spatially selective deposition of plasmonic NPs in a laser interference pattern. It opens the possibility of 'on-site' creation of nano-grating structures formed of periodically distributed arrays of NPs. The sensitivity of the deposits' morphology to the parameters of the LID process is demonstration of its flexibility and potency for creation of fine structures for tailoring light-matter interaction and sensing technologies development.

2. Materials and Methods

2.1. Materials and Reagents

Analytical grade solvents (dichlorethane, dichlormethane, isopropanol, ethanol, methanol, acetonitrile, hexane) were purchased from Reachem (Reachem, Moscow, Russia), and used after standard purification procedures [27]. In addition, for solutions preparation, double-distilled water was used. Cover glass and quartz slips with 0.15 mm thickness and 5 mm × 5 mm size (Levenhuk G100 cover slips; model 5932-020-00288679-2012) (Tampa, FL, USA) were used as substrates for LID. As a source of metal, we used two silver-containing precursors and two platinum-containing precursors: silver benzoate hydrate (silver salt C_6H_5COOAg, containing of Ag 47.1 wt%), silver bis(trifluoromethylsulfonyl)azanide (silver salt, $(CF_3SO_2)_2NAg$ containing of Ag 27.6 wt%), ethenyl-[ethenyl(dimethyl)silyl]oxy-dimethylsilane platinum in vinyl terminated polydimethylsiloxane 0.1 M solution (platinum complex $C_8H_{18}OPtSi_2$, containing of Pt 3 wt%), (1Z,5Z)-cycloocta-1,5-diene dichloroplatinum (platinum salt $(C_8H_{12})Cl_2Pt$, containing of Pt 51.5 wt%). Precursors were purchased from Alfa Aesar and used as received. The molecular structures of these substances are presented in Figure S1 Supporting Information.

2.2. Laser-Induced Deposition

The compositions of the precursors' solutions for the laser-induced deposition are presented in Table 1. Preparation of the solutions was carried out with ultrasonic treatment for 5 min and subsequent centrifugation at 12,000 rpm for 3 min in a Sigma 2–16P (Sigplma Laborzentrifugen, Osterode am Harz, Germany). Because of low solubility of $(C_8H_{12})Cl_2Pt$ in C_6H_{14} and H_2O and low miscibility of vinyl terminated polydimethylsiloxane with H_2O, these solutions were not used for further LID experiments.

Table 1. Concentrations (mM) of precursors in solutions used for laser-induced deposition (LID).

Solvent	Precursor			
	C_6H_5COOAg	$(CF_3SO_2)_2NAg$	$C_8H_{18}OPtSi_2$	$(C_8H_{12})Cl_2Pt$
$C_2H_4Cl_2$	4.5	2.6	1.3	1.3
CH_3OH	0.4	0.8	26.2	0.2
C_6H_{14}	2.2	4.2	3.9	-
C_3H_8O	1.1	2.8	13.1	3.0
H_2O	1.7	0.5	-	-
C_2H_3N	1.3	1.3	3.9	2.7
CH_2Cl_2	1.3	0.8	2.9	5.3
C_2H_5OH	0.9	3.3	3.9	4.8

All the prepared solutions were considered for laser-induced deposition of monometallic (Pt and Ag) structures; the binary Pt-Ag system was deposited from mixed acetonitrile solution of C_6H_5COOAg (7.5 mM) and $C_8H_{18}OPtSi_2$ in vinyl terminated polydimethylsiloxane (3.9 mM). The absorption spectra of all the studied solutions are presented in Supplementary Materials (Figure S2). The laser wavelength for the LID process was chosen in accordance with characteristic absorption bands of the studied precursors. That is why a solid-state, continuous wave, single frequency, deep-UV laser system Coherent MBD266 (wavelength 266 nm, power 60 mW, unfocused laser beam, laser spot diameter 2 mm) was used as a radiation source for LID realization. The laser parameters were kept the same for all LID experiments.

The scheme of the LID process is presented in Figure 1a. Unfocused laser beam with diameter ca. 2 mm was directed to the substrate–solution interface through the solution. The volume of the cuvette was 80 µL, thickness of the solution 1 mm. LID was carried out in a stationary regime—no shift of laser beam relative to the substrate. Laser irradiation time was 40 min in all the experiments. In the case of platinum complexes, LID process was performed both on quartz substrates and cover glass slips, deposition from silver complexes was performed on cover glass slips. After the LID process, the substrates were washed with isopropanol and dried at ambient conditions.

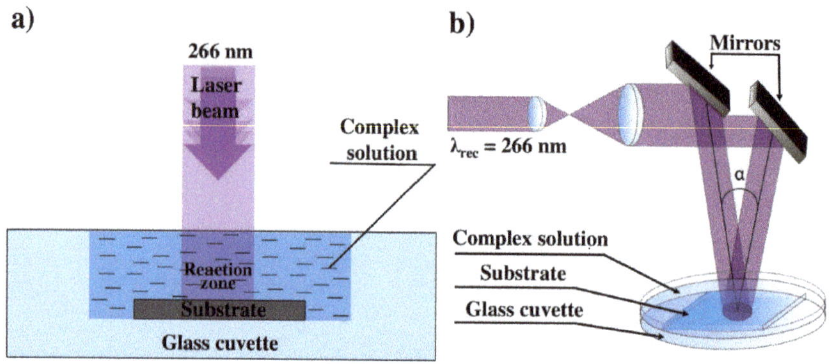

Figure 1. (**a**) Reaction system of LID; (**b**) Experimental setup for synthesis of periodic silver nanoparticle structures using holographic recording.

To investigate the effect of laser intensity on the LID process, the formation of deposits in the laser interference pattern was studied. A regular two-beam scheme was used for the interference pattern construction [28]. The scheme of the experimental setup is shown in Figure 1b. Two parallel laser beams of equal intensities were produced by dividing the input laser beam that was expanded and collimated before. These two coherent beams were directed at an angle α to each other to form an interference pattern. The period d of the interference pattern was determined by the formula $2dsin(\alpha/2) = \lambda_{rec}$, where λ_{rec} is a laser wavelength. The LID process was performed for two interference patterns—with 375 and 750 lines/mm. The LID process in the interference pattern was carried out similarly to the LID in a single laser beam. As a deposition solution in the interference pattern, we used 0.3 mM silver benzoate hydrate C_6H_5COOAg in methanol. The LID process in the interference pattern was carried out for laser power 45 and 200 mW and a range of irradiation time from 2 to 20 min, the irradiated area was ca. 0.5 cm^2.

2.3. Samples Characterization

The absorption spectra of solutions were recorded with SHIMADZU UV-2550 over the spectral range 200–500 nm with 1 nm step, medium scan speed, and 1 nm slit width. The absorption spectra of NPs deposited on cover glass or quartz slip were recorded with a Lambda 1050 (Perkin Elmer, Waltham, MA, USA) in the range of 300–550 nm (for silver NPs) and 200–550 nm (for platinum NPs) with an integrating Ulbricht sphere. The substrate with NPs was placed in the sphere center. The morphology and composition of obtained samples were investigated by scanning electron microscopy (SEM) and energy-dispersive X-ray spectroscopy (EDX) with a scanning electron microscope Zeiss Merlin (Oberkochen, Germany) with field emission cathode, GEMINI electron-optics column, oil-free vacuum system, using variable pressure charge compensation mode with local nitrogen injection. No conductive coating was used. Crystallinity and composition of the samples were investigated by transmission electron microscopy (TEM) with Zeiss Libra 200. For TEM investigation, the samples were sonicated in isopropanol and deposited onto the carbon-coated TEM grid. The crystal structure was also studied by X-ray diffractometer Bruker "D8 DISCOVER" (Billerica, MA, USA), using Cu Kα X-ray line. Atomic force microscopy (AFM) (NT-MDT, Moscow, Russia) analysis of deposits obtained under interference pattern was performed with NTEGRA-Prima setup in tapping mode in areas of 2×60 µm and 14×14 µm with 256 points/line.

3. Results and Discussion

3.1. Characterization of Monometallic Ag, Pt Nanostructures Obtained by LID

Laser-induced deposition for studied Pt and Ag precursors was found to be successful for all the solutions listed in Table 1. The result of the LID process was the formation of nanostructures in the laser-affected area of the substrate. The density and homogeneity of the nanostructure distribution on the substrate surface exhibited some spatial variation depending on the precursor and the solvent. The most typical deposits are presented in Figures 2 and 3.

Figure 2 demonstrates the SEM images of the structures obtained by the LID procedure from silver precursors (C_6H_5COOAg and $(CF_3SO_2)_2NAg$). The LID process results in formation of nanoparticles that are homogeneously distributed on the substrate's surface. The NPs are characterized by a faceted morphology in the case of C_6H_5COOAg precursor with rather wide size distribution: 50–200 nm for the methanol solvent and 100–500 nm for acetonitrile. The larger size of NPs from C_6H_5COOAg solution in acetonitrile is most likely determined by higher concentration of C_6H_5COOAg in acetonitrile than in methanol, as all other deposition parameters were kept identical.

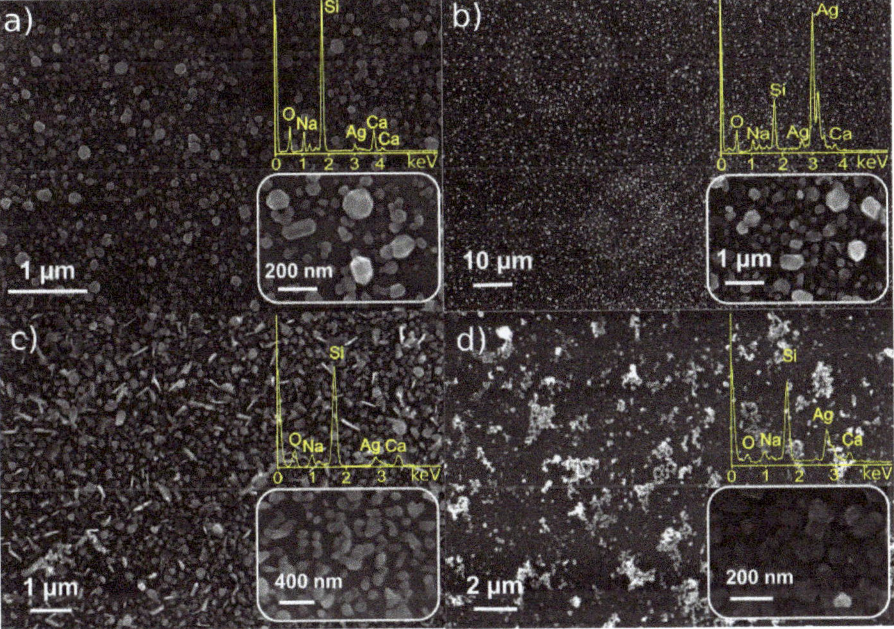

Figure 2. SEM/EDX analysis of Ag particles on cover glass synthesized from (**a**) C_6H_5COOAg in methanol (0.4 mM), (**b**) C_6H_5COOAg in acetonitrile (1.3 mM), (**c**) $(CF_3SO_2)_2NAg$ in ethanol (3.3 mM), (**d**) $(CF_3SO_2)_2NAg$ in isopropanol (2.8 mM).

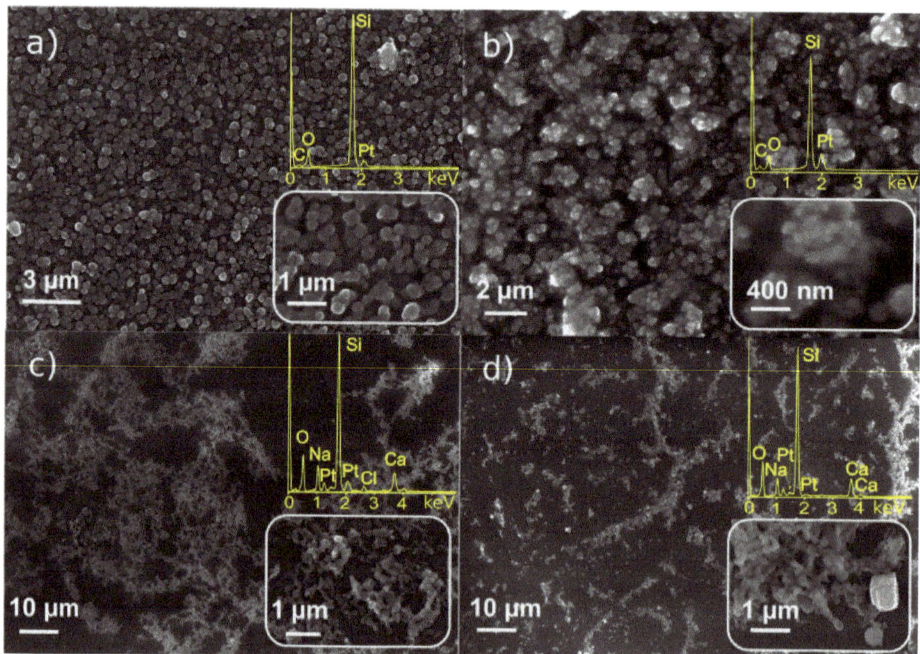

Figure 3. SEM/EDX analysis of particles synthesized from (**a**) $C_8H_{18}OPtSi_2$ in vinyl terminated polydimethylsiloxane (0.1 M), (**b**) $C_8H_{18}OPtSi_2$ in hexane C_6H_{14} (3.9 mM), (**c**) $(C_8H_{12})Cl_2Pt$ in dichloromethane CH_2Cl_2 (5.3 mM), (**d**) $(C_8H_{12})Cl_2Pt$ in dichlorethane $C_2H_4Cl_2$ (1.3 mM).

The LID from $(CF_3SO_2)_2NAg$ precursor also results in homogeneously distributed NPs; in the case of $(CF_3SO_2)_2NAg$ in isopropanol, the layer of NPs is thicker. The NPs do not reveal faceted morphology—$(CF_3SO_2)_2NAg$ solution in ethanol results in formation of platelets, while $(CF_3SO_2)_2NAg$ in isopropanol demonstrates aggregated NPs; the average size is about 150 nm for both solvents. EDX analysis testifies that all the deposited NPs consist of silver; the observed O, Na, Si, Ca are signals from the substrates (cover slips), no other impurities were found.

The SEM images of nanostructures obtained from platinum precursors are presented in Figure 3. The deposition from $C_8H_{18}OPtSi_2$ precursor demonstrates formation of NPs with highly homogeneous distribution onto the substrate's quartz surface (Figure 3a,b). Dilution with hexane of "native" precursor solution in vinyl terminated polydimethylsiloxane results in the formation of less continuous coatings consisting of agglomerated NPs.

Figure 3c,d presents SEM images of nanostructures deposited from $(C_8H_{12})Cl_2Pt$ precursor. Generally, the deposits do not form homogeneous coating as distinct from the $C_8H_{18}OPtSi_2$ complex (Figure 3a,b). The morphology of structures synthesized from solutions of $(C_8H_{12})Cl_2Pt$ in dichloroethane and dichloromethane is similar; the size of particles is close, but in case of dichloromethane, we see a higher coverage. EDX analysis testifies the presence of Pt in all the deposited samples, however in the case of $C_8H_{18}OPtSi_2$ precursor, one can see C and O signals that probably originate from the precursor complex. O, Na, Si, Ca peaks in Figure 3c,d are due to the signal from the microscope cover slip substrate.

Thus, SEM and EDX analysis of the structures obtained by LID technique demonstrates that the most homogeneous and regular structures consisted of Pt and Ag nanoparticles can be obtained correspondingly from $C_8H_{18}OPtSi_2$ and C_6H_5COOAg precursors. Variations of the solvent have pronounced effects, and enable the experimentalist to adjust the morphology of the deposit. For the further experiments on LID synthesis of the binary Pt-Ag system, the mixed solution of $C_8H_{18}OPtSi_2$ and C_6H_5COOAg precursors in acetonitrile was chosen.

3.2. Characterization of Binary Pt-Ag System Obtained by LID

LID from the mixed solution of the $C_8H_{18}OPtSi_2$ and C_6H_5COOAg precursors in acetonitrile resulted in the formation of a continuous coating on the substrate surface (Figure 4a). One can see a rather homogeneous distribution of 20–30 nm NPs with several larger particles of 200–400 nm that are aggregates of small NPs. In spite of similarity in the morphologies of monometallic (Ag, Pt) and binary Pt-Ag NPs, the characterization of the latter requires more detailed analysis and additional techniques to uncover the composition and structure of the deposited phase.

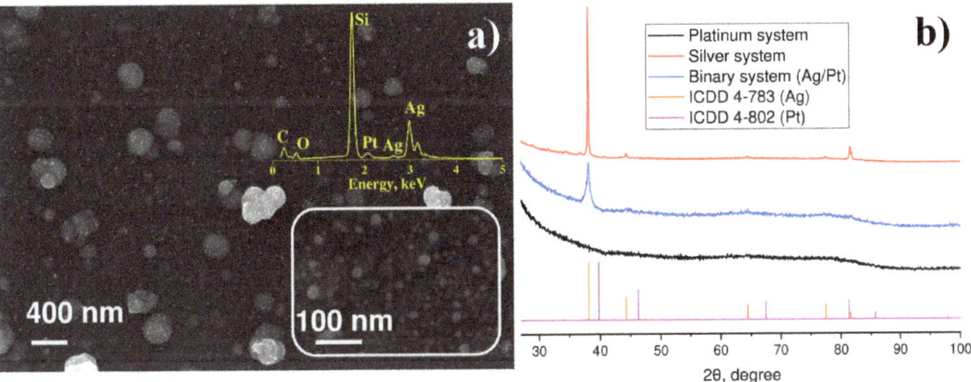

Figure 4. (a) SEM/EDX analysis of the of particles synthesized from mixed Pt-Ag precursors, (b) X-ray diffraction of particles on quartz substrates for platinum system $C_8H_{18}OPtSi_2$; silver system (C_6H_5COOAg in acetonitrile); binary system ($C_8H_{18}OPtSi_2$+ C_6H_5COOAg in acetonitrile); ICDD PDF-2 for platinum (4-802 card) and silver (4-783 card).

Figure 4b presents experimental XRD patterns of nanoparticles obtained from C_6H_5COOAg and $C_8H_{18}OPtSi_2$ complexes, respectively, their mixture, and the PDF-2 database cards for silver and platinum. One can see that Ag NPs demonstrated pronounced diffraction peaks, which coincide well with the silver card ICDD 4-783 for both monometallic and binary systems. NPs obtained from $C_8H_{18}OPtSi_2$ complexes did not show any trace of diffraction peaks and displayed only diffuse halo. This fact can be explained by formation of amorphous platinum NPs or small size of Pt crystallites. To determine the correct reason, further investigation of synthesized samples was carried out by means of electron microscopy and microanalysis.

TEM analysis of deposits from C_6H_5COOAg precursor demonstrate the formation of rather large crystalline particles, with diameters on the order of 100–200 nm (Figure 5a), that consist of silver in accordance with EDX microanalysis (Figure 5b). Copper and iron peaks are attributed to the sample grid and specimen holder. The electron diffraction pattern (Figure 5c) consists of several rings, as typical for randomly oriented crystals. Measuring the diameters of these rings, we have obtained the interplanar spacings of 2.40 Å, 2.08 Å, 1.46 Å, and 1.22 Å, which are identical within experimental uncertainty to literature values for the spacings in face-centered cubic lattice of silver (2.36 Å, 2.04 Å, 1.44 Å, and 1.18 Å) [29]. Thus, both electron diffraction and EDX microanalysis (Figure 5b,c) unambiguously show that these particles consist of silver only.

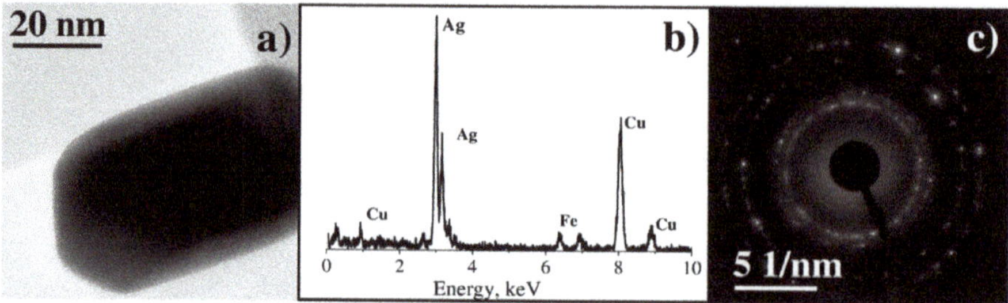

Figure 5. (**a**) Bright-field (BF) TEM image of large single crystal, (**b**) EDX spectrum, (**c**) electron diffraction pattern.

As one can observe from the BF TEM image (Figure 6a), the sample deposited from the $C_8H_{18}OPtSi_2$ precursor consists of crystalline nanoparticles in an amorphous matrix. EDX analysis of single particle demonstrates several peaks in the spectrum, which correspond to the C Kα line, O Kα line, Pt Mα line, and Si Kα line (Figure 6b, blue spectrum). The spectrum measured from an amorphous matrix (Figure 6b, red spectrum) includes C Kα line, O Kα line, and Si Kα line. Thus, it can be concluded that the sample consists of Pt nanoparticles (ca. 1–3 nm in diameter) incorporated into an amorphous matrix, consisting of carbon, silicon, and oxygen. The origin of the amorphous matrix is the decomposition of ligand in the precursor's complex $C_8H_{18}OPtSi_2$. Several rings with diameters corresponding to the interplanar spacings of 2.29 Å, 1.98 Å, 1.41 Å, and 1.20 Å are observed in electron diffraction (Figure 6c), which can be attributed to crystalline Pt (2.27 Å, 1.96 Å, 1.39 Å, 1.18 Å) [30].

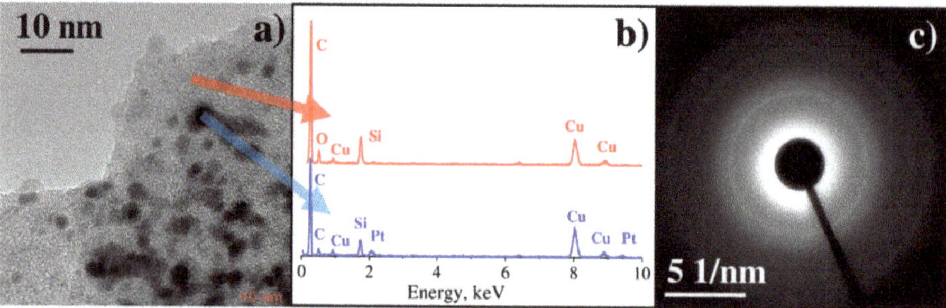

Figure 6. (**a**) BF TEM image, (**b**) EDX spectrum from single nanoparticle and spectrum from the matrix, (**c**) electron diffraction pattern.

TEM investigation of the sample deposited from the mixture of C_6H_5COOAg and $C_8H_{18}OPtSi_2$ complexes shows that it consists of two types of nanoparticles: small ones with the size of several nanometers, and large ones with the size of hundreds of nanometers (Figure 7a).

The EDX spectrum measured from small nanoparticles (Figure 7b, blue spectrum) exhibits an X-ray peak at 2.05 keV that corresponds to the Pt Mα line. An EDX spectrum measured from large nanoparticles (Figure 7b, green spectrum) shows a prominent peak at 2.98 keV that corresponds to the Ag Lα line. Electron diffraction consists of several rings and some bright reflexes (Figure 7c). The reflexes correspond to the interplanar spacings: 2.40 Å, 2.08 Å, 1.46 Å, and 1.22 Å, typical for crystalline silver, whereas rings correspond to 2.29 Å, 1.98 Å, 1.41 Å, and 1.20 Å, typical for crystalline platinum.

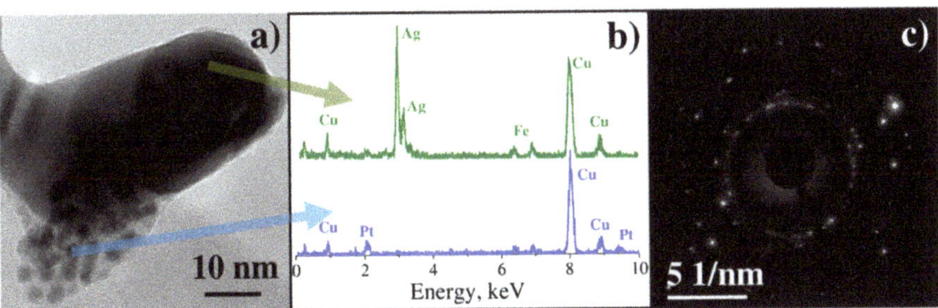

Figure 7. (**a**) Bright-field (BF) TEM image of sample 1-2.1, (**b**) X-ray spectrum from small nanoparticles and spectrum from large nanoparticles, (**c**) electron diffraction pattern.

Therefore, the detailed analysis performed with SEM, TEM, EDX, and X-ray diffraction techniques unambiguously revealed the formation of ca. 100 nm crystalline nanoparticles of silver as a result of the LID process from C_6H_5COOAg and $(CF_3SO_2)_2NAg$ precursors, while Pt deposits are characterized by smaller size 1–5 nm. It should be noted that despite the absence of specific X-ray reflexes corresponding to Pt crystals in the diffraction patterns of Pt and mixed Pt-Ag samples, TEM electron diffraction gives unambiguous confirmation of crystalline nature of Pt deposits.

It should be noted that the detailed description of the mechanism of formation of Ag and Pt NPs from the solutions of precursors studied here require additional investigation. Nevertheless, keeping in mind the resonance absorption of laser radiation by precursor molecules and the absence of reducing agent as a component of the solution, one can conclude that laser excitation of the precursor molecules is followed by their decomposition/transformation and intramolecular redox processes resulting in formation of metal phase. In the case of the $C_8H_{18}OPtSi_2$ precursor, we found ligand incorporation into the structure of deposit along with Pt clusters. LID from the binary Ag-Pt solutions demonstrated independent processes of formation of Ag and Pt phases resulting in precipitation of mixture of NPs.

3.3. LID-Obtained Structures for Optical Application

The nanostructures deposited in the current study can be considered as prospective material for application in a range of fields, for example, surface-enhanced Raman spectroscopy (SERS) or ultrasensitive optical sensors based on plasmonic enhancement of electromagnetic fields. Plasmonic properties of metal nanoparticles reveal in UV-VIS absorption spectra as characteristic bands, which positions depend on metal nature [31], size, and shape of NPs [32], and refractive index of the media [33]. In case of multimetallic NPs, the characteristic absorption bands are sensitive to the nature of metal phase (alloy or mechanical mixture) [34,35]. Thus, absorption spectroscopy can be considered as a sensitive tool for characterization of both physicochemical properties of NPs and their functionality.

Figure 8 shows absorption spectra of NPs deposited as a result of the LID process from solutions of C_6H_5COOAg, $(CF_3SO_2)_2NAg$, $C_8H_{18}OPtSi_2$, and $(C_8H_{12})Cl_2Pt$ precursors. Rather broad bands in range 375–475 nm observed for NPs from Ag-containing precursors (Figure 8a) correspond to the plasmon-related absorption of silver nanostructures and testify to a wide size distribution of NPs. The faceted elongated shape of silver NPs confirmed by SEM can also contribute to the plasmonic spectra width [36].

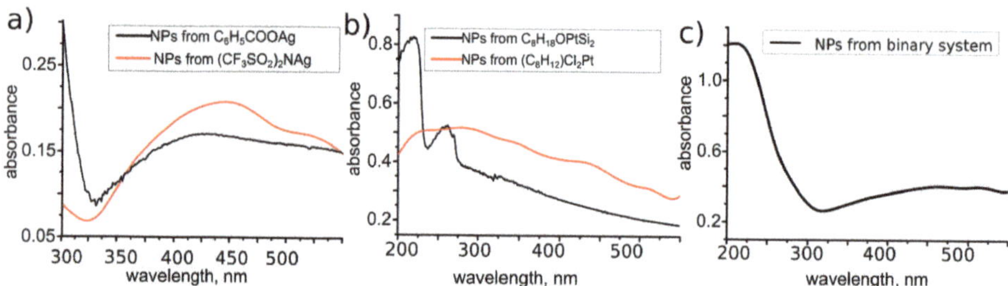

Figure 8. Absorption spectra of nanoparticles (NPs) deposited from (**a**) C_6H_5COOAg in methanol and $(CF_3SO_2)_2NAg$ in isopropanol; (**b**) $C_8H_{18}OPtSi_2$ in vinyl terminated polydimethylsiloxane and $(C_8H_{12})Cl_2Pt$ in dichloromethane; (**c**) mixture of C_6H_5COOAg and $C_8H_{18}OPtSi_2$ in acetonitrile.

Absorption spectra of NPs from Pt-containing precursors (Figure 8b) demonstrate more specific features, especially for the solution of $C_8H_{18}OPtSi_2$. One can see pronounced peaks at 220 and 260 nm, which are attributed to the intra-band electronic transitions from energy states of ($n = 5; l = 2$) and ($n = 6; l = 0$) to higher energy states of the conduction bands of Pt NPs [37,38]. Pt NPs obtained from $(C_8H_{12})Cl_2Pt$ precursor display a wide absorption band centered at about 270 nm. The width of the band may be caused by overlapping of peaks related with faceted shape of NPs and broad sizes distribution. Difference of absorption spectra of Pt NPs deposited from various precursors could be due to size and shape variation.

It is interesting to note that absorption spectra of deposits obtained from the mixed solution of $C_8H_{18}OPtSi_2$ and C_6H_5COOAg precursors display two separated peaks with maxima 220 and 450 nm. The peaks positions are typical for Pt and Ag nanoparticles, respectively, and their presence confirms the formation of a binary system consisting of a mixture of Pt and Ag NPs. These data coincide well with the described above TEM analysis and absorption spectra for monometallic NPs.

The peculiarity of the LID presented here is the use of low intensity laser irradiation and laser wavelengths corresponding to the characteristic absorption of the precursor, all together proving the photo-induced nature of NPs' formation process. In this case, the deposition process can be sensitive to the spatial variation of laser intensity and promising for the controlled NPs distribution on the substrate surface. As a proof of concept, the LID process was realized in an interference pattern for the C_6H_5COOAg precursor in methanol. Figure 9 shows an optical micrograph and SEM images of Ag NPs gratings obtained in different synthesis conditions (laser power, deposition time, diffraction period). One can see high sensitivity of the deposits' morphology on the synthesis parameters. Deposition at high laser power (200 mW) and short laser irradiation time (2 min) generates periodic arrays with and without NPs (Figure 9a,b), but its quality is not high enough for demonstration of diffraction. Decreasing the laser power while keeping short illumination time does not result in formation of a periodic pattern of NPs. Further optimization of deposition parameters results in a full coverage of the substrate surface with NPs, at the same time, the periodicity of the morphology is clearly observed (Figure 9c–e). It is interesting to note that the average NPs' size for the 'line' and 'gap' areas is close (ca. 30 nm) but the size distribution is narrower in case of 'gap' areas (S3). Figure 9f demonstrates diffraction of the He–Ne laser beam (633 nm, 5 mW) on the sample, presented in Figure 9e.

The morphology of the deposited gratings was studied with AFM analysis. AFM images of the cross-section and surface of the sample (laser power of 45 mW, irradiation time 40 min, interference pattern 375 lines/mm) are given in Figure 10. An average thickness of the sample of ca. 40 nm was obtained as a result of a scan through a scratch line. Figure 10b confirms the formation of a surface relief grating with height (difference between minima and maxima) approximately 25 nm and a period ca 2.6 μm, according to Figure 10c. The results obtained coincide perfectly with the interference pattern period that was 2.67 μm for the sample studied here.

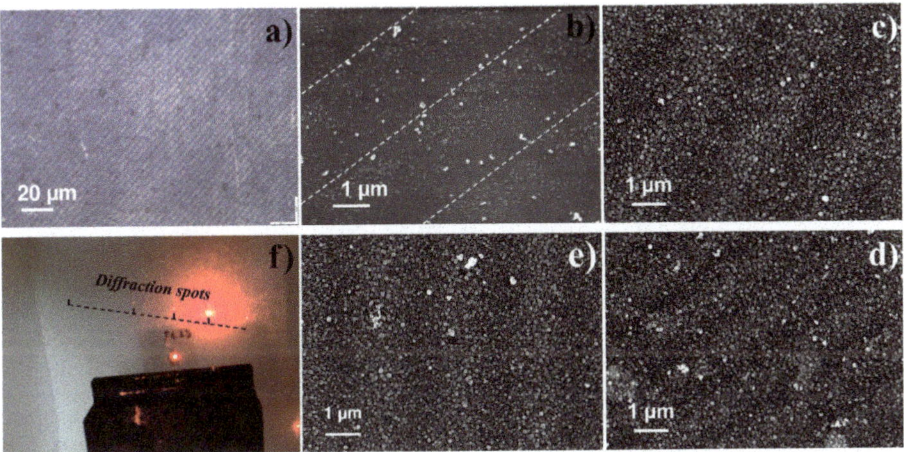

Figure 9. Micrograph and SEM analysis of Ag NPs grating with different synthesis conditions: (**a,b**) laser power of 200 mW, irradiation time 2 min, period 375 lines/mm; (**c**) laser power of 45 mW, irradiation time 20 min, period 375 lines/mm; (**d**) laser power of 45 mW, irradiation time 40 min, period 750 lines/mm; (**e**) laser power of 200 mW, irradiation time 10 min, period 375 lines/mm; (**f**) demonstration of diffraction effect on Ag NPs grating deposited under laser power of 200 mW, irradiation time 10 min, period 375 lines/mm.

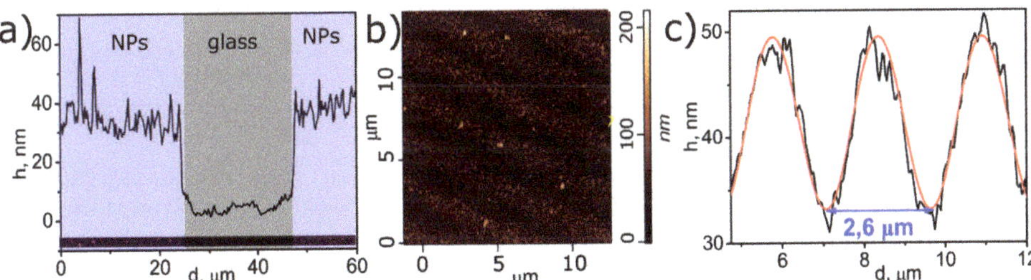

Figure 10. (**a**) AFM image and average profile of LID-synthesized sample; (**b**) AFM image of Ag NPs grating deposited under laser beam with interference pattern 375 lines/mm; (**c**) average profile (black line) and fitting waveform (red line) of synthesized periodic structure.

4. Conclusions

In this work, we successfully demonstrated the formation of Ag, Pt, and mixed Ag-Pt nanoparticles on the surface of glass and quartz slips as a result of the LID process from solutions of commercially available precursors—C_6H_5COOAg, $(CF_3SO_2)_2NAg$, $C_8H_{18}OPtSi_2$, and $(C_8H_{12})Cl_2Pt$. The deposition process was found to be possible from solutions of all the studied precursors in a wide range of solvents. A pronounced effect of solvent on the deposits' morphology was observed which can be used to deposit NPs with desired properties. The deposited NPs were characterized with SEM, TEM, EDX, X-ray analysis. In the case of deposition from the solutions containing one type of precursors, formation of crystalline platinum NPs with average size 1–5 nm, and silver nanoparticles ca. 100 nm was found. Deposition from the binary Ag-Pt solutions revealed formation of a mixture of monometallic crystalline Ag and Pt NPs that testifies toward independent processes of Ag and Pt phase formation during LID. X-ray analysis revealed good correlation of Ag reflexes with database, while Pt reflexes were not detected. At the same time, electron diffraction patterns of the studied samples demonstrated interplanar spacings typical for crystalline silver and crystalline platinum. The obtained X-ray results show that absence of crystalline features in XRD patterns does not necessary testify amorphous phase, independent investigation with electron diffraction can be more informative towards small crystallites detection.

UV-VIS absorption spectroscopy demonstrated plasmonic absorption of all the studied samples and gave independent confirmation of formation of NPs mixture in the case of the Ag-Pt sample. It should be noted that the mechanism of NPs formation from the solutions of the studied precursors requires further understanding; however, resonance absorption of laser radiation by precursor molecules and the following intramolecular redox processes resulting in formation of metal nano-species can be considered as a general description of the LID process.

The unique feature of the LID procedure presented is the use of low-intensity laser irradiation that provides the photo-induced nature of the deposition process, the absence of the destructive effect of laser radiation on the deposits, and the potential for precise control of the deposits' spatial distribution. As a proof of concept, the spatially selective deposition of plasmonic NPs in the laser interference pattern was successfully demonstrated. Diffraction gratings with different parameters 375 lines/mm and 750 lines/mm formed of periodically distributed Ag NPs were created and characterized. The obtained results open a wide range of opportunities for the LID process. Firstly, the list of potential precursors can be extended towards commercially available precursors to make LID widely accessible. Secondly, a number of applications can be explored where different kinds of plasmonic effects are utilized (tailoring light, non-invasive optical detection methods, sensing technologies).

Supplementary Materials: The following are available online at https://www.mdpi.com/1996-194 4/14/1/10/s1, Figure S1: Structures of LID precursors, Figure S2: Absorbance spectra of precursors in different solvents (a) C7H5AgO2; (b) C2AgF6NO4S2; (c) O[Si(CH3)2 CH=CH2]2Pt; (d) C8H12Cl2Pt; Figure S3: (a) Particles size histogram for line area of Ag NPs grating; (b) SEM for line and gap areas of Ag NPs grating, laser power of 200 mW, irradiation time 20 min periods of the interference pattern 3 m (c) Particles size histogram for gap area of Ag NPs grating.

Author Contributions: Conceptualization, D.V.M. and A.A.V.; methodology, Y.V.P.; investigation, D.V.M., A.A.V. and D.V.D.; data curation, Y.V.P.; writing—original draft preparation, D.V.M., A.A.V. and I.E.K.; writing—review and editing, A.A.M. and J.B.; visualization, A.A.K.; supervision, A.A.M. and J.B.; project administration, A.A.M.; funding acquisition, A.A.M. All authors have read and agreed to the published version of the manuscript.

Funding: This work was supported by joint RFBR-DFG project (RFBR project No 20-58-12015, DFG project BA 4277/16-1) (platinum and bimetallic part); RFBR project No 19-33-90239 (silver part), by the "Scholarships of the President of the Russian Federation to young scientists and graduate students (Competition SP-2019)", project number СП-2368.2019.1. Authors are grateful to "Centre for Optical and Laser materials research", "Interdisciplinary Resource Centre for Nanotechnology", "Physics Educational Centre" and "Centre for X-ray Diffraction Studies" of Research Park of Saint Petersburg State University for technical support.

Informed Consent Statement: Not applicable.

Data Availability Statement: Please refer to suggested Data Availability Statements in section "MDPI Research Data Policies" at https://www.mdpi.com/ethics. Studies" of Research Park of Saint Petersburg State University for technical support.

Conflicts of Interest: The authors reported no conflict of interest related to this study.

References

1. Lin, F.; Yang, J.; Lu, S.-H.; Niu, K.-Y.; Liu, Y.; Sun, J.; Du, X.-W. Laser synthesis of gold/oxide nanocomposites. *J. Mater. Chem.* **2010**, *20*, 1103–1106. [CrossRef]
2. Marzun, G.; Levish, A.; Mackert, V.; Kallio, T.; Barcikowski, S.; Wagener, P. Laser synthesis, structure and chemical properties of colloidal nickel-molybdenum nanoparticles for the substitution of noble metals in heterogeneous catalysis. *J. Colloid Interface Sci.* **2017**, *489*, 57–67. [CrossRef] [PubMed]
3. Yin, H.; Zhao, Y.; Xu, X.; Chen, J.; Wang, X.; Yu, J.; Wang, J.; Wu, W. Realization of Tunable Localized Surface Plasmon Resonance of Cu@Cu2O Core–Shell Nanoparticles by the Pulse Laser Deposition Method. *ACS Omega* **2019**, *4*, 14404–14410. [CrossRef] [PubMed]
4. Nakamura, T.; Magara, H.; Herbani, Y.; Ito, A.; Sato, S. Fabrication of gold-platinum nanoparticles by intense, femtosecond laser irradiation of aqueous solution. In Proceedings of the Conference on Lasers and Electro-Optics/International Quantum Electronics Conference, Munich, Germany, 14–19 June 2009; Paper JWA2. Optical Society of America: Washington, DC, USA, 2009; p. JWA2.
5. Nakamura, T.; Herbani, Y.; Sato, S. Fabrication of solid-solution gold-platinum nanoparticles with controllable compositions by high-intensity laser irradiation of solution. *J. Nanopart. Res.* **2012**, *14*, 785. [CrossRef]
6. Guay, J.-M.; Calà Lesina, A.; Côté, G.; Charron, M.; Poitras, D.; Ramunno, L.; Berini, P.; Weck, A. Laser-induced plasmonic colours on metals. *Nat. Commun.* **2017**, *8*, 16095. [CrossRef] [PubMed]
7. Mikšys, J.; Arutinov, G.; Römer, G.R.B.E. Pico- to nanosecond pulsed laser-induced forward transfer (LIFT) of silver nanoparticle inks: A comparative study. *Appl. Phys. A* **2019**, *125*, 814. [CrossRef]
8. Li, C.; Hu, J.; Jiang, L.; Xu, C.; Li, X.; Gao, Y.; Qu, L. Shaped femtosecond laser induced photoreduction for highly controllable Au nanoparticles based on localized field enhancement and their SERS applications. *Nanophotonics* **2020**, *9*, 691–702. [CrossRef]
9. Manshina, A.; Povolotskiy, A.; Ivanova, T.; Kurochkin, A.; Tver'yanovich, Y.; Kim, D.; Kim, M.; Kwon, S.C. Laser-assisted metal deposition from CuSO4-based electrolyte solution. *Laser Phys. Lett.* **2007**, *4*, 163–167. [CrossRef]
10. Seo, J.M.; Kwon, K.-K.; Song, K.Y.; Chu, C.N.; Ahn, S.-H. Deposition of Durable Micro Copper Patterns into Glass by Combining Laser-Induced Backside Wet Etching and Laser-Induced Chemical Liquid Phase Deposition Methods. *Materials* **2020**, *13*, 2977. [CrossRef]
11. Peng, C.; Richard Liu, C.; Voothaluru, R.; Ou, C.-Y.; Liu, Z. An Exploratory Investigation of the Mechanical Properties of the Nanostructured Porous Materials Deposited by Laser-Induced Chemical Solution Synthesis. *J. Micro Nanomanuf.* **2017**, *5*. [CrossRef]
12. Bjerneld, E.J.; Murty, K.V.G.K.; Prikulis, J.; Käll, M. Laser-Induced Growth of Ag Nanoparticles from Aqueous Solutions. *ChemPhysChem* **2002**, *3*, 116–119. [CrossRef]
13. Bjerneld, E.J.; Svedberg, F.; Käll, M. Laser-Induced Growth and Deposition of Noble-Metal Nanoparticles for Surface-Enhanced Raman Scattering. *Nano Lett.* **2003**, *3*, 593–596. [CrossRef]
14. Zhang, Z.; Xiao, X.; Yang, C. Laser-induced pattened deposition of silver nanoparticles on glass. In Proceedings of the International Conference on Advanced Infocomm Technology 2011 (ICAIT 2011), Wuhan, China, 11–14 July 2011; p. 26. [CrossRef]
15. Ming-Shan, L.; Chang-Xi, Y. Laser-Induced Silver Nanoparticles Deposited on Optical Fiber Core for Surface-Enhanced Raman Scattering. *Chin. Phys. Lett.* **2010**, *27*, 044202. [CrossRef]
16. Meng, X.; Bi, Z.; Shang, G. Laser-induced Ag nanoparticles deposition on optical fiber probes for TERS. In Proceedings of the the the International Photonics and Optoelectronics Meeting (POEM), Wuhan, China, 31 October–3 November 2018; Paper OT4A.74. Optical Society of America: Washington, DC, USA, 2018; p. OT4A.74.
17. Povolotskiy, A.; Povolotckaia, A.; Petrov, Y.; Manshina, A.; Tunik, S. Laser-induced synthesis of metallic silver-gold nanoparticles encapsulated in carbon nanospheres for surface-enhanced Raman spectroscopy and toxins detection. *Appl. Phys. Lett.* **2013**, *103*, 113102. [CrossRef]
18. Vasileva, A.; Haschke, S.; Mikhailovskii, V.; Gitlina, A.; Bachmann, J.; Man'shina, A. Direct laser-induced deposition of AgPt@C nanoparticles on 2D and 3D substrates for electrocatalytic glucose oxidation. *Nano-Struct. Nano-Objects* **2020**, *24*, 100547. [CrossRef]
19. Manshina, A.; Povolotskiy, A.; Ivanova, T.; Kurochkin, A.; Tver'yanovich, Y.; Kim, D.; Kim, M.; Kwon, S.C. CuCl2-based liquid electrolyte precursor for laser-induced metal deposition. *Laser Phys. Lett.* **2006**, *4*, 242. [CrossRef]
20. Manshina, A. Laser-Induced Deposition of Metal and Hybrid Metal-Carbon Nanostructures. In *Progress in Photon Science: Recent Advances*; Yamanouchi, K., Tunik, S., Makarov, V., Eds.; Springer Series in Chemical Physics; Springer International Publishing: Cham, Switzerland, 2019; pp. 387–403. ISBN 978-3-030-05974-3.
21. Bashouti, M.Y.; Manshina, A.; Povolotckaia, A.; Povolotskiy, A.; Kireev, A.; Petrov, Y.; Mačković, M.; Spiecker, E.; Koshevoy, I.; Tunik, S.; et al. Direct laser writing of μ-chips based on hybrid C-Au-Ag nanoparticles for express analysis of hazardous and biological substances. *Lab Chip* **2015**, *15*, 1742–1747. [CrossRef]
22. Bashouti, M.Y.; Povolotckaia, A.V.; Povolotskiy, A.V.; Tunik, S.P.; Christiansen, S.H.; Leuchs, G.; Manshina, A.A. Spatially-controlled laser-induced decoration of 2D and 3D substrates with plasmonic nanoparticles. *RSC Adv.* **2016**, *6*, 75681–75685. [CrossRef]

23. Schlicht, S.; Kireev, A.; Vasileva, A.; Grachova, E.V.; Tunik, S.P.; Manshina, A.A.; Bachmann, J. A model electrode of well-defined geometry prepared by direct laser-induced decoration of nanoporous templates with Au-Ag@C nanoparticles. *Nanotechnology* **2017**, *28*. [CrossRef]
24. Kolesnikov, I.E.; Lvanova, T.Y.; Ivanov, D.A.; Kireev, A.A.; Mamonova, D.V.; Golyeva, E.V.; Mikhailov, M.D.; Manshina, A.A. In-situ laser-induced synthesis of associated YVO4:Eu3+@SiO2@Au-Ag/C nanohybrids with enhanced luminescence. *J. Solid State Chem.* **2018**, *258*, 835–840. [CrossRef]
25. Haschke, S.; Pankin, D.; Mikhailovskii, V.; Barr, M.K.S.; Both-Engel, A.; Manshina, A.; Bachmann, J. Nanoporous water oxidation electrodes with a low loading of laser-deposited Ru/C exhibit enhanced corrosion stability. *Beilstein J. Nanotechnol.* **2019**, *10*, 157–167. [CrossRef] [PubMed]
26. Povolotckaia, A.; Pankin, D.; Petrov, Y.; Vasileva, A.; Kolesnikov, I.; Sarau, G.; Christiansen, S.; Leuchs, G.; Manshina, A. Plasmonic carbon nanohybrids from laser-induced deposition: Controlled synthesis and SERS properties. *J. Mater. Sci.* **2019**, *54*, 8177–8186. [CrossRef]
27. Armarego, W.L.F.; Chai, C. *Purification of Laboratory Chemicals*, 6th ed.; Elsevier: Oxford, UK, 2009; ISBN 978-1-85617-567-8.
28. Smirnova, T.N.; Kokhtych, L.M.; Kutsenko, A.S.; Sakhno, O.V.; Stumpe, J. The fabrication of periodic polymer/silver nanoparticle structures: In situ reduction of silver nanoparticles from precursor spatially distributed in polymer using holographic exposure. *Nanotechnology* **2009**, *20*, 405301. [CrossRef] [PubMed]
29. Spreadborough, J.; Christian, J.W. High-temperature X-ray diffractometer. *J. Sci. Instrum.* **1959**, *36*, 116–118. [CrossRef]
30. Herbani, Y.; Nakamura, T.; Sato, S. Synthesis of platinum-based binary and ternary alloy nanoparticles in an intense laser field. *J. Colloid Interface Sci.* **2012**, *375*, 78–87. [CrossRef] [PubMed]
31. Zheng, W.; Chiamori, H.C.; Liu, G.L.; Lin, L.; Chen, F.F. Nanofabricated plasmonic nano-bio hybrid structures in biomedical detection. *Nanotechnol. Rev.* **2012**, *1*, 213–233. [CrossRef]
32. Seo, D.; Park, J.C.; Song, H. Polyhedral Gold Nanocrystals with Oh Symmetry: From Octahedra to Cubes. *J. Am. Chem. Soc.* **2006**, *128*, 14863–14870. [CrossRef]
33. Pastoriza–Santos, I.; Sánchez–Iglesias, A.; García de Abajo, F.J.; Liz–Marzán, L.M. Environmental Optical Sensitivity of Gold Nanodecahedra. *Adv. Funct. Mater.* **2007**, *17*, 1443–1450. [CrossRef]
34. Rubio, A.S. *Modified Au-Based Nanomaterials Studied by Surface Plasmon Resonance Spectroscopy*; Springer Theses; Springer International Publishing: Berlin/Heidelberg, Germany, 2015; ISBN 978-3-319-19401-1.
35. Zaleska-Medynska, A.; Marchelek, M.; Diak, M.; Grabowska, E. Noble metal-based bimetallic nanoparticles: The effect of the structure on the optical, catalytic and photocatalytic properties. *Adv. Colloid Interface Sci.* **2016**, *229*, 80–107. [CrossRef]
36. Petryayeva, E.; Krull, U.J. Localized surface plasmon resonance: Nanostructures, bioassays and biosensing—A review. *Anal. Chim. Acta* **2011**, *706*, 8–24. [CrossRef]
37. Gharibshahi, E.; Saion, E. Influence of Dose on Particle Size and Optical Properties of Colloidal Platinum Nanoparticles. *Int. J. Mol. Sci.* **2012**, *13*, 14723–14741. [CrossRef] [PubMed]
38. Gharibshahi, E.; Saion, E.; Johnston, R.L.; Ashraf, A. Theory and experiment of optical absorption of platinum nanoparticles synthesized by gamma radiation. *Appl. Radiat. Isot.* **2019**, *147*, 204–210. [CrossRef] [PubMed]

 *materials*

Article

Laser-Induced Synthesis of Composite Materials Based on Iridium, Gold and Platinum for Non-Enzymatic Glucose Sensing

Maxim S. Panov [1], Evgeniia M. Khairullina [1], Filipp S. Vshivtcev [2], Mikhail N. Ryazantsev [1,2,*] and Ilya I. Tumkin [1,*]

[1] Saint Petersburg State University, 7/9 Universitetskaya Nab., 199034 St. Petersburg, Russia; m.s.panov@spbu.ru (M.S.P.); e.khayrullina@spbu.ru (E.M.K.)

[2] Nanotechnology Research and Education Centre of the Russian Academy of Sciences, Saint Petersburg Academic University, 194021 St. Petersburg, Russia; vforvshivcev@gmail.com

* Correspondence: mikhail.n.ryazantsev@gmail.com (M.N.R.); i.i.tumkin@spbu.ru (I.I.T.)

Received: 23 June 2020; Accepted: 24 July 2020; Published: 29 July 2020

Abstract: A simple approach for in situ laser-induced modification of iridium-based materials to increase their electrocatalytic activity towards enzyme-free glucose sensing was proposed. For this purpose, we deposited gold and platinum separately and as a mixture on the surface of pre-synthesized iridium microstructures upon laser irradiation at a wavelength of 532 nm. Then, we carried out the comparative investigation of their morphology, elemental and phase composition as well as their electrochemical properties. The best morphology and, as a result, the highest sensitivity ($\sim$9960 μA/mM cm^2) with respect to non-enzymatic determination of D-glucose were demonstrated by iridium-gold-platinum microstructures also showing low limit of detection ($\sim$0.12 μM), a wide linear range (0.5 μM–1 mM) along with good selectivity, reproducibility and stability.

Keywords: laser-induced deposition of metal microstructures; iridium; gold; platinum; composites; enzyme-free sensing; glucose

---

1. Introduction

There are many methods for determination of biologically significant compounds such as hydrogen peroxide, glucose, amino acids and many others [1,2]. Methods based on usage of the electrochemical sensors are the most useful and effective among them [3,4]. Electrochemical sensors are devices designed for qualitative and quantitative analysis of liquid and gaseous media, in which the analytical signal is provided by the flow of electrons emerging due to an electrochemical reaction in the near-electrode space. The central feature of any sensors that distinguishes them from other analytical devices is ability to conduct field analysis in real time and minimal sample preparation. Electrochemical sensing and monitoring of glucose as one of the most important substrates for the diagnosis and treatment of diabetes attract particular attention in this regard [4]. The main principle of operation of the most well-known glucose meters and sensors is an electrochemical method for measuring glucose concentration based on its enzymatic cleavage. Such measurements proceed through two stages. At the first stage, glucose from the blood ends up in the test area and under the action of the enzyme glucose oxidase is broken down into gluconic acid and hydrogen peroxide [3,5,6]. At the second stage, each hydrogen peroxide molecule under the action of a small electric field decays to form an oxygen molecule, two protons and two electrons. Further, each glucose molecule gives off two electrons creating an electric current, which, in turn, is detected by the analyzer. Nevertheless, it is known that these enzyme sensors for glucose detection have a number of disadvantages: low accuracy of detection, low stability due to enzyme decomposition and high sensitivity to pH and humidity of the environment leading to a short service life of such devices [6]. The reason for this is the interfering influence of biologically active substances present in the blood as well as the type of materials used as electrodes in these sensors. Indeed, for example, thin-film electrodes based on gold or platinum with immobilized glucose oxidase are frequently used, in which activity of an enzyme decreases very quickly over time, thereby reducing the lifetime of such sensors. Moreover, the methods used for production of thin-film electrodes (e.g., vacuum deposition) do not provide high reproducibility, so a separate calibration is required for each sensor, which, in turn, affects the accuracy of determining the glucose concentration in the blood [6–12]. In opposite, our approach proposed avoids such disadvantages and assumes a more selective and accurate invasive biochemical blood analysis that can be performed both in the laboratory and at home. This approach is based on the use of laser-induced metal deposition from solution (LCLD) to produce enzyme-free electrochemical microsensors suitable for glucose detection [13]. In LCLD, the metal reduction reaction occurs in a local volume of a solution within the focus of the laser beam leading to its deposition on the surface of a dielectric substrate (e.g., glass, glass-ceramics etc.). This method allows the synthesis of electrocatalytically active materials based on different metals and has a number of advantages over many known analogues due to its simplicity, low cost and sufficient efficiency [14–21]. The sensors synthesized by LCLD can be applied for direct detection of biologically important analytes. It is known that enzymatic detection of such analytes is complicated by high values of their potentials. In contrast, in non-enzymatic detection, the value of these potentials can be reduced due to highly developed structure of the electrode materials, which can also be modified with gold or platinum micro- or nanoscaled particles resulting in catalysis of ox-red reactions of analytes (e.g., glucose) on the surface of an electrode [22–24]. As a result, non-enzymatic sensing combines the advantages of the electrochemical methods and the unique properties of nanomaterials exhibiting large surface area due to three-dimensional structures of such nanoparticles of catalytically active and biologically compatible metals as gold and platinum. This leads to significant decrease of the analysis time and increase of sensitivity, stability and intensity of the analytical signal in the determination of glucose concentration.

In this work, we synthesized mono- and polymetallic microstructures based on iridium, gold and platinum using LCLD technique. We also performed comparative study of their morphology, composition and electrochemical properties. Iridium-containing metal and oxide micro- and nanostructures demonstrated rather good electrocatalytic activity towards pH sensing and enzyme-free detection of many biologically important substrates [25–28]. This makes them a good choice for application in non-enzymatic glucose sensing.

2. Materials and Methods

Analytically graded chloro(triphenylphosphine)gold(I), tris(2-phenylpyridine)iridium(III), dichloro(dicyclopentadienyl)platinum(II), *N,N*-dimethylformamide (DMF) were used as received from Sigma Aldrich (St. Louis, MO, USA) without any further purification. All solutions were prepared with deionized water with a resistivity of greater than 18.1 MΩ·cm^{-1}. The compositions of solutions for LCLD are shown in Table 1. Silica-based glass-ceramics slides were used as substrate for laser-induced deposition process.

Table 1. The compositions of solutions used for LCLD experiments.

Component	Concentration (mM)	Solvent
tris(2-phenylpyridine)iridium(III)	3	DMF
chloro(triphenylphosphine)gold(I)	3	DMF
dichloro(dicyclopentadienyl)platinum(II)	3	DMF

The scheme of the experimental setup and its detailed description can be found in Supplementary Materials (Figure S1) and elsewhere [13]. An neodymium-doped yttrium aluminum garnet (Nd:YAG) diode-pumped solid-state cw laser (Changchun New Industries Optoelectronics Technology Co., Ltd.,

Changchun, China) with a wavelength of 532 nm was used as a light source for the thermo-induced reduction and deposition of metals. The output from a laser is split into two portions. First portion of the laser output provides in situ monitoring the process of laser-induced synthesis using web-camera. In turn, the second portion is used to govern reduction reaction. Microdeposits were produced by scanning the laser beam focused between the plating solution and dielectric substrate using an XYZ controlled motorized stage. Consequently, we were able to deposit the 8 mm long and 70–90 μm wide iridium (Ir) lines at the laser power of 1500 mW and at the scanning speed of 2.5 μm s^{-1}. Finally, the synthesized Ir lines were modified by the consecutive laser-induced deposition of gold and platinum on the top of them using the same deposition rate and at laser power of 1100 mW.

Scanning electron microscopy (SEM) images of the synthesized iridium containing microdeposits were obtained with Zeiss Supra 40 VP scanning electron microscope (Oberkochen, Germany). Energy-dispersive X-ray spectroscopy (EDX) was used to identify atomic composition of microelectrodes, all samples were studied using an INCA X-Act EDX analyzer (Oxford Instruments, Abingdon, UK) coupled with SEM.

XRD measurements were carried out on Bruker D2 Phaser diffractometer equipped with LynxEye detector (Karlsruhe, Germany) using CuKα (0.1542 nm) radiation in the 2θ angle range of 0–100°.

The electronic absorption spectra of the DMF solutions of triphenylphosphine chloride of gold(I), dicyclopentadienyl platinum(II) dichloride and tris(2-phenylpyridine)iridium(III) were recorded using UV-vis spectrophotometer Shimadzu UV-2550 (Duisburg, Germany).

Electrochemical characterization was performed in a conventional three-electrode cell at ambient temperature, using Elins P30I potentiostat (Electrochemical Instruments Ltd., Chernogolovka, Russia). Platinum wire, Ag/AgCl electrode and Ir-based microelectrodes was used as counter, reference and working electrodes respectively. All electrochemical measurements were done in 0.1 M sodium hydroxide saturated with Ar as background solution. The electrocatalytic activity of the fabricated Ir-based materials towards glucose was studied using CV and CA techniques. Cyclic voltammetry (CV) experiments were executed at a scan rate of 50 mV·s^{-1} between −0.8 and 0.8 V vs. Ag/AgCl. CA was implement by the addition of the solutions of D-glucose of different concentrations to background solution with simultaneous stirring.

3. Results and Discussion

The microstructures based on iridium, gold and platinum were synthesized using the method of laser-induced deposition of metals from solution (LCLD). Thus, we produced iridium (Ir) microdeposits from a solution of tris(2-phenylpyridine)iridium(III) in *N,N*-dimethylformamide (DMF) on the surface of glass-ceramics. Iridium-platinum (Ir-Pt) and iridium-gold (Ir-Au) microstructures were obtained by consecutive laser-induced deposition of platinum and gold from DMF solutions of dicyclopentadienyl platinum(II) dichloride and triphenylphosphine chloride of gold(I), respectively, on the surface of pre-synthesized iridium microdeposits. Iridium-gold-platinum (Ir-Au-Pt) microstructures were produced by consecutive laser-induced deposition of platinum on the surface of the pre-synthesized iridium-gold (Ir-Au).

In fact, photochemical reactions potentially can contribute in decomposition of metal complexes used for LCLD in this study. Photochemistry of such complexes is defined by the photochemical activity of their excited states associated with the ligand-to-metal charge transfer (LMCT). In general, the electronic absorption spectra of metal complexes reveal intense LMCT bands in the UV-vis region as well as absorption bands located at longer wavelengths (including near-IR range) assigned to d-d transitions. The excitation to the d-d states does not induce the photoreaction, whereas the LMCT excited states are dissociative and lead to the reduction of metal complex followed by formation of metallic structures [29–32]. In turn, the 532 nm excitation used in the current work does not populate the photochemically active LMCT states of 1, 2 and 3 (Figure 1) meaning that in our case we observe thermo-induced metal reduction process.

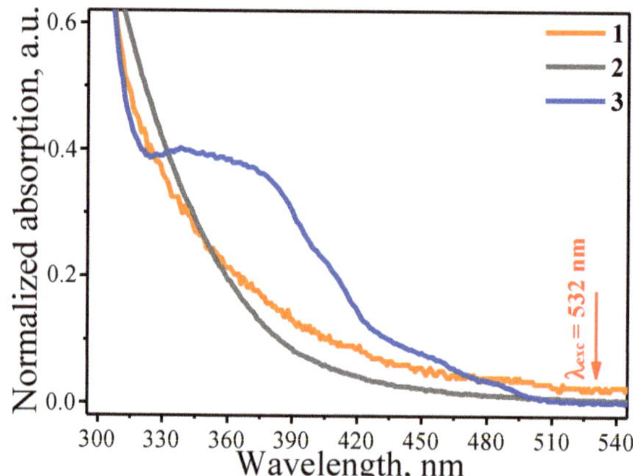

Figure 1. UV-vis absorption spectra normalized to unity at the absorption maximum of the DMF solutions of triphenylphosphine chloride of gold(I) (1), dicyclopentadienyl platinum(II) dichloride (2) and tris(2-phenylpyridine)iridium(III) (3).

Figure 2 demonstrates the morphology of these Ir-based materials examined using scanning electron microscopy (SEM). The surface of the fabricated iridium microdeposits has continuous and uniform structure (Figure 2a–c). Modification of the surface of iridium with platinum did not lead to a significant increase in surface development. SEM revealed that platinum forms micro-sized structures of irregular shape on the surface of iridium (Figure 2d–f). Iridium microdeposits doped with gold have both insular and continuous structure, the surface of which is more developed in comparison with two previous materials (Figure 2g–i). Herein, gold is deposited as separate micro- and nanoscaled drops (Figure 2i), which are prone to agglomeration (Figure 2h). The most interesting results were obtained by successive deposition of platinum on the surface of Ir-Au. Ir-Au-Pt microstructures are similar to Ir-Au in a number of points, but have several significant differences (Figure 2j–l). These microdeposits have highly developed surface consisting of spherical particles with a diameter of 40–400 nm of gold and platinum, both individually and as a mixture (Figure 2l). Unlike Ir-Au, there is no complete agglomeration of small particles into clusters, but there is a significant variation in their size (Figure 2k,l). The results of energy-dispersive X-ray (EDX) and X-ray diffraction (XRD) analysis are shown in Figure 3 and Supplementary Materials (Figures S2–S7). EDX-analysis exhibited the presence of some other elements (besides iridium, gold and platinum) that are most likely associated with the substrate material and components of the plating solutions. XRD of the synthesized microstructures showed the presence of a polyphase multicomponent mixture containing metal and oxide phases (Figure 3). For all materials, the metal phase is iridium, whereas, the oxide phase is iridium dioxide (IrO_2). Additional metal phases were observed for Ir-Pt-platinum and for Ir-Au-gold. The presence of IrO_2 in all microdeposits is consistent with their sufficiently high values of electrical resistance (~0.4–1.5 kΩ), which are close to those of semiconductors.

Figure 2. SEM images of (**a–c**) iridium (Ir), (**d–f**) iridium-platinum (Ir-Pt), (**g–i**) iridium-gold (Ir-Au) and (**j–l**) iridium-gold-platinum (Ir-Au-Pt) microstructures.

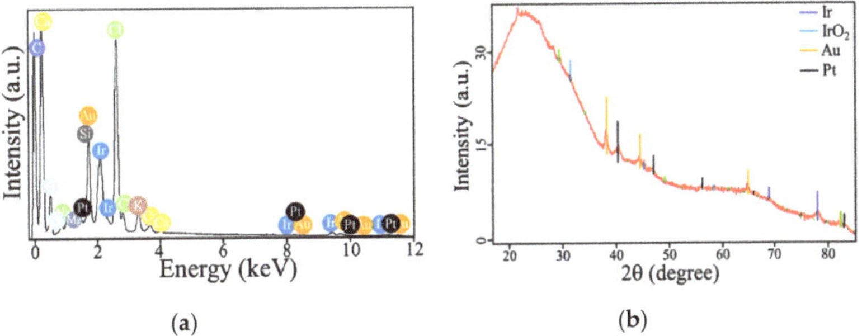

Figure 3. EDX spectrum (**a**) and XRD pattern (**b**) of iridium-gold-platinum (Ir-Au-Pt) microstructures.

Figure 4a illustrates cyclic voltammograms (CVs) of all types of iridium-containing microstructures recorded in solutions of 1 mM D-glucose. Here, it should be noted that the area and shape of CV curve of Ir-Au-Pt differs significantly from other Ir-based microelectrodes. It is known that the area of a CV is directly related to the degree of development of the surface of an electrode. Moreover, the value of this parameter determines the sensitivity of an electrode with respect to a particular analyte. As shown in Figure 4a, modification of the iridium (Ir) microdeposits with platinum does not considerably improve its sensory properties. In opposite, modification of iridium with gold and, in particular, with a mixture of gold and platinum causes a noticeable increase in the area of CV curve. As a result, consecutive laser-induced deposition of iridium, gold and platinum allows producing a microelectrode with a much higher analytical response. Considering this, our further electrochemical studies were focused only on Ir-Au-Pt. Figure 4b presents cyclic voltamperograms of Ir-Au-Pt microstructures recorded in a background solution (0.1 M NaOH) with different concentration of D-glucose. The peaks of potentials corresponding to electrocatalytic oxidation and reduction of glucose for Ir-Au-Pt have a pronounced character for the entire region of analyte concentrations. The region of anodic glucose oxidation is located approximately between −100 and 100 mV, moreover, there are two regions of cathodic oxidation of glucose lying between −500 and −300 mV and between 0 and 100 mV. The second region of cathodic glucose oxidation shifts with an increase of the concentration of D-glucose (Figure 4b).

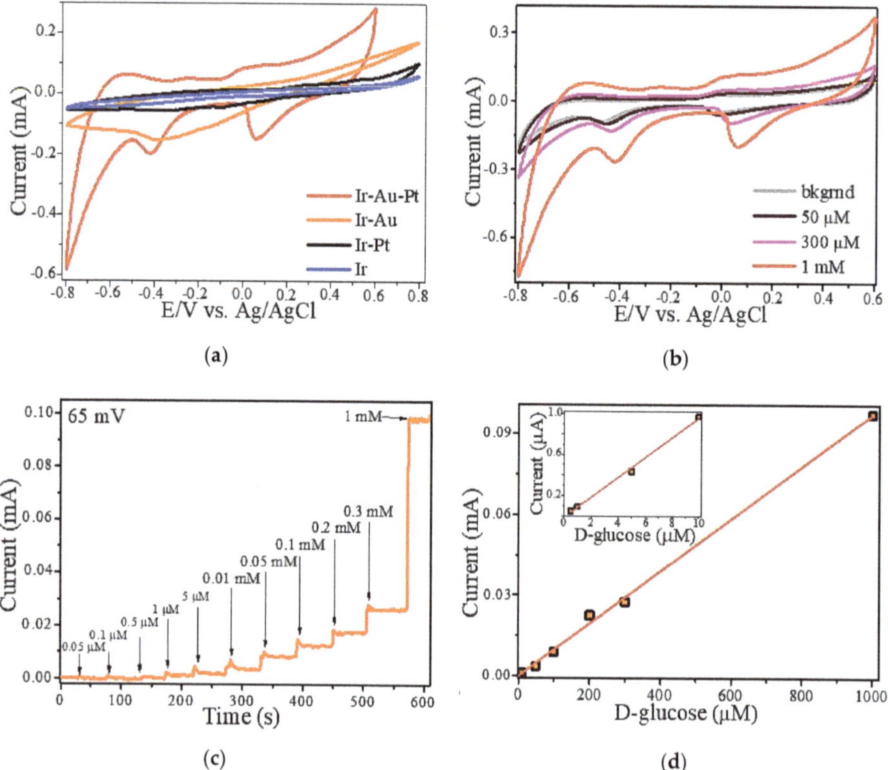

Figure 4. Cyclic voltammograms (CVs) of microstructures based on iridium, gold and platinum recorded in 0.1 M NaOH containing 1 mM D-glucose (**a**). CV of Ir-Au-Pt recorded in 0.1 M NaOH with different concentration of D-glucose (**b**). Amperogram of Ir-Au-Pt obtained in 0.1 M NaOH with different concentration of D-glucose (**c**). Linear dependence of the measured Faraday current on the D-glucose concentrations (**d**).

In addition, the amperometric method was also used to evaluate the electrocatalytic activity of Ir-Au-Pt microelectrode towards glucose detection. Figure 4c illustrates a typical amperogram that characterizes the reaction of Ir-Au-Pt to successive additions of D-glucose to the background solution at a potential of 65 mV corresponding to the region of the double electric layer [24]. It is clear that an increase of the concentration of D-glucose leads to an increase in the analytical signal. The linear dependence of the Faraday current for Ir-Au-Pt measured at a potential of 65 mV on concentration of D-glucose is shown in Figure 4d. It can be seen that Ir-Au-Pt demonstrates a linear range of enzyme-free glucose determination lying between 0.5 µM and 1 mM. The limit of detection (LOD) of glucose for

Ir-Au-Pt microelectrode was calculated as LOD = 3 S/b, where S is the standard deviation from linearity and b is the slope of the calibration curve (linear range shown in Figure 4d). Thus, the calculated value of LOD for Ir-Au-Pt equals to ~0.12 μM and the maximum estimated sensitivity, which is proportional to the area of CV curve shown in Figure 4a, is ~9960 μA/mM cm^2. The low detection limit and high sensitivity exhibited by Ir-Au-Pt can be explained by the large surface area of this material and the electrocatalytic synergy between iridium, gold and platinum. It is known that bimetallic and trimetallic Pt-based catalysts are desirable due to the presence of an additional metal that can enhance the catalytic activity via synergy because of the electronic, alloying or strain effects. In turn, all component of the alloy catalysts can affect the activity enhancement [33]. Moreover, not only the electrocatalytic activity, but also the stability of bimetallic and trimetallic noble metal nanocrystals (NCs) are higher than those of their monometallic counterparts due to the modification of their electronic structures, available catalytically active sites and other positive synergistic effects caused by structural and compositional differences between the monometallic and bimetallic NCs [34–36]. However, the mechanisms that lead to a synergistic effect and an increase of catalytic activity towards glucose electrooxidation are quite complex and individual for each composite; therefore, in order to clarify such mechanism for the material synthesized in this work the separate study is required. In addition, it is thought that active transition metal centers across the electrode and the presence of hydroxyl radicals play a crucial role in electrooxidation of glucose and many other organic molecules [24,37,38]. Therefore, taking into account the models known from literature and mentioned above, we can assume that catalytic oxidation of glucose on the surface of the synthesized Ir-Au-Pt microelectrode may proceed via the mechanism demonstrated in Figure 5.

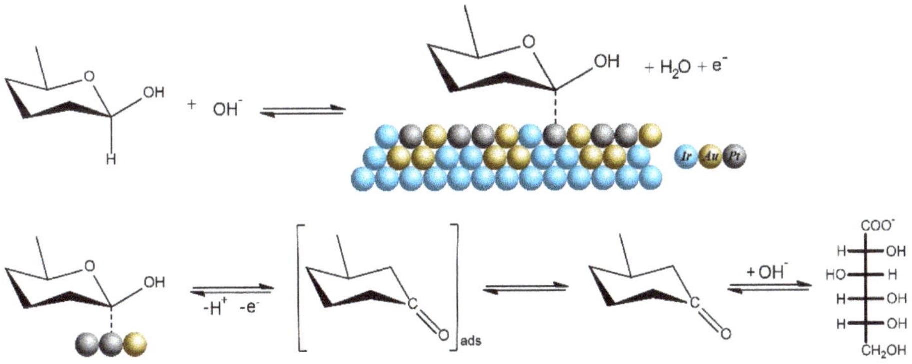

Figure 5. Schematic representation of a possible mechanism for oxidation of glucose at Ir-Au-Pt microelectrode.

We have also tested the selectivity of the Ir-Au-Pt with respect to D-glucose in the presence of several interfering substances including ascorbic acid (AA), 4-acetamidophenol (AP), uric acid (UA) and hydrogen peroxide (H$_2$O$_2$) that typically coexist with glucose in the human blood (Figure 6). Figure 6 shows that addition of these substances to the background solution (0.1 M NaOH) leads to an increase of the amperometric current at the applied potential of 65 mV. The most prominent analytical response in comparison with other analytes was observed for D-glucose. Thus, according to these observations, it can be concluded that Ir-Au-Pt exhibits rather decent selectivity towards enzyme-free glucose sensing.

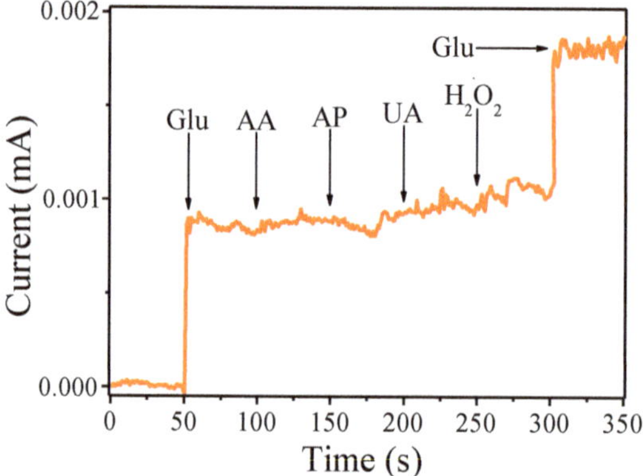

Figure 6. Amperometric response of Ir-Au-Pt microelectrode to successive addition of 50 μM D-glucose (Glu), 10 μM ascorbic acid (AA), 10 μM 4-acetamidophenol (AP), 10 μM uric acid (UA), 10 μM hydrogen peroxide (H$_2$O$_2$) in 0.1 M NaOH.

Furthermore, rather good electrochemical characteristics of iridium microelectrode modified with gold and platinum are accompanied with decent stability and reproducibility. Long-term stability was investigated with three Ir-Au-Pt microelectrodes for two weeks. As a result, it was observed that this material maintained ~91% its electrocatalytic activity with respect to non-enzymatic detection of D-glucose during the first week, whereas this activity dropped to ~85% during the second week. The great reproducibility was confirmed by low value of the relative standard deviation related to electrochemical response to 0.1 mM solution of D-glucose, which was found to be around 10% for three Ir-Au-Pt electrodes.

In summary, sensor characteristics of Ir-Au-Pt microelectrode were compared with same characteristics of non-enzymatic sensors based on similar metal and bimetallic micro- and nanostructures known from literature [7,27,28,39–45]. It should be noted that this microelectrode has a fairly low detection limit, a wide range of linearity and high sensitivity in relation to enzyme-free glucose sensing in comparison with many modern analogues (Table 2).

Table 2. Comparison of different electrode materials for non-enzymatic glucose sensing.

Electrode Material	Sensitivity (μA/mM × cm^2)	Linear Range (mM)	Limit of Detection (μM)	Ref.
This work (Ir-Au-Pt)	9960	0.0005–1	0.12	-
IrO$_2$ NFs Nafion/GCE	22.22	0–16	2.93	[28]
Ir-carbon	48.83	0–50	28,000	[39]
Pt-Ir	93.7	0–10	N/A	[27]
Nanoporous Pt	10.0	1–10	50	[40]
Pt NP	22.7	0–10	0.42	[44]
Pd-Cu-Pt	553	1–10	1.29	[41]
Pt-Au nanoporous film	12.85	0.2–4.8	1.3	[45]
Pt-nanoporous Au	145.7	0.5–10	0.6	[43]
Au nanocorals	22.6	0.0005–0.3	10	[7]
Au NP film	749.2	0.0556–13.89	9	[42]

4. Conclusions

In this work, we used laser-induced metal deposition technique in order to manufacture several iridium-containing microelectrodes. It was found that modification of iridium electrode with gold and platinum both individually and as a mixture leads to better surface development and, as a result, enhances electrocatalytic activity towards non-enzymatic glucose sensing. In this regard, the most promising results were demonstrated by the composite microstructures based on iridium, gold and platinum exhibiting great electrochemical properties. It can also be noted that this material is of great interest for the design and development of inexpensive and effective portable sensor devices for enzyme-free determination of various biologically important analytes.

Supplementary Materials: The following are available online at http://www.mdpi.com/1996-1944/13/15/3359/s1, Figure S1: The experimental setup used for laser-induced deposition of iridium-containing microstructures: 1—cw 532 nm diode-pumped solid-state Nd:YAG laser, 2—mirrors, 3—splitting cube, 4—objective lens, 5—the experimental cell, 6—neutral density flter, 7—web camera, 8—personal computer, 9—the computer controlled motorized stage, Figure S2: EDX spectrum of iridium (I_r) microstructures, Figure S3: EDX spectrum of iridium-platinum (Ir-Pt) microstructures, Figure S4: EDX spectrum of iridium-gold (Ir-Au) microstructures; Figure S5: XRD pattern of iridium (Ir) microstructures, Figure S6: XRD pattern of iridium-platinum (Ir-Pt) microstructures, Figure S7: XRD pattern of iridium-gold (Ir-Au) microstructures.

Author Contributions: Conceptualization and writing, M.S.P.; investigation and data curation, E.M.K.; synthesis and formal analysis, F.S.V.; supervision and writing—review and editing, M.N.R.; project administration and funding acquisition, I.I.T. All authors have read and agreed to the published version of the manuscript.

Funding: This research was partially supported by Fellowship of President of Russia MK-1521.2020.3. I.I.T., M.S.P. and E.M.K. acknowledge the financial support by RFBR, project number 20-33-70277. M.N.R. and F.S.V. acknowledge the financial support from The Ministry of Education and Science of Russian Federation (project number FSRM-2020-0006).

Acknowledgments: The authors express their gratitude to the SPbSU Nanotechnology Interdisciplinary Centre, Centre for Optical and Laser Materials Research and Centre for X-ray Diffraction Studies.

Conflicts of Interest: The authors declare no conflict of interest.

References

1. Hahn, Y.B.; Ahmad, R.; Tripathy, N. Chemical and biological sensors based on metal oxide nanostructures. *Chem. Commun.* **2012**, *48*, 10369–10385. [CrossRef]
2. Trial, C. Effect of intensive therapy on the development and progression of diabetic nephropathy in the Diabetes Control and Complications Trial. *Kidney Int.* **1995**, *47*, 1703–1720. [CrossRef]
3. Bakker, E. Electrochemical sensors. *Anal. Chem.* **2004**, *76*, 3285–3298. [CrossRef]
4. Saei, A.A.; Dolatabadi, J.E.N.; Najafi-Marandi, P.; Abhari, A.; de la Guardia, M. Electrochemical biosensors for glucose based on metal nanoparticles. *TrAC Trends Anal. Chem.* **2013**, *42*, 216–227. [CrossRef]
5. Heikenfeld, J.; Jajack, A.; Rogers, J.; Gutruf, P.; Tian, L.; Pan, T.; Li, R.; Khine, M.; Kim, J.; Wang, J.; et al. Wearable sensors: Modalities, challenges, and prospects. *Lab. Chip* **2018**, *18*, 217–248. [CrossRef]
6. Ferri, S.; Kojima, K.; Sode, K. Review of glucose oxidases and glucose dehydrogenases: A bird's eye view of glucose sensing enzymes. *J. Diabetes Sci. Technol.* **2011**, *5*, 1068–1076. [CrossRef]
7. Cheng, T.M.; Huang, T.K.; Lin, H.K.; Tung, S.P.; Chen, Y.L.; Lee, C.Y.; Chiu, H.T. (110)-Exposed gold nanocoral electrode as low onset potential selective glucose sensor. *ACS Appl. Mater. Interfaces* **2010**, *2*, 2773–2780. [CrossRef] [PubMed]
8. Lay, B.; Coyle, V.E.; Kandjani, A.E.; Amin, M.H.; Sabri, Y.M.; Bhargava, S.K. Nickel-gold bimetallic monolayer colloidal crystals fabricated: Via galvanic replacement as a highly sensitive electrochemical sensor. *J. Mater. Chem. B* **2017**, *5*, 5441–5449. [CrossRef]
9. Li, C.; Wang, H.; Yamauchi, Y. Electrochemical deposition of mesoporous Pt-Au alloy films in aqueous surfactant solutions: Towards a highly sensitive amperometric glucose sensor. *Chem.-A Eur. J.* **2013**, *19*, 2242–2246. [CrossRef]
10. Xu, C.; Liu, Y.; Su, F.; Liu, A.; Qiu, H. Nanoporous PtAg and PtCu alloys with hollow ligaments for enhanced electrocatalysis and glucose biosensing. *Biosens. Bioelectron.* **2011**, *27*, 160–166. [CrossRef]
11. Mizoshiri, M.; Kondo, Y. Direct writing of two- and three-dimensional Cu-based microstructures by femtosecond laser reductive sintering of the Cu_2O nanospheres. *Opt. Mater. Express* **2019**, *9*, 2828–2837. [CrossRef]
12. Kwon, J.; Cho, H.; Suh, Y.D.; Lee, J.; Lee, H.; Jung, J.; Kim, D.; Lee, D.; Hong, S.; Ko, S.H. Flexible and Transparent Cu Electronics by Low-Temperature Acid-Assisted Laser Processing of Cu Nanoparticles. *Adv. Mater. Technol.* **2017**, *2*, 1600222. [CrossRef]

13. Kochemirovsky, V.A.; Skripkin, M.Y.; Tveryanovich, Y.S.; Mereshchenko, A.S.; Gorbunov, A.O.; Panov, M.S.; Tumkin, I.I.; Safonov, S.V. Laser-induced copper deposition from aqueous and aqueous–organic solutions: State of the art and prospects of research. *Russ. Chem. Rev.* **2015**, *84*, 1059–1075. [CrossRef]

14. Panov, M.S.; Tumkin, I.I.; Mironov, V.S.; Khairullina, E.M.; Smikhovskaia, A.V.; Ermakov, S.S.; Kochemirovsky, V.A. Sensory properties of copper microstructures deposited from water-based solution upon laser irradiation at 532 nm. *Opt. Quantum Electron.* **2016**, *48*, 490. [CrossRef]

15. Baranauskaite, V.E.; Novomlinskii, M.O.; Tumkin, I.I.; Khairullina, E.M.; Mereshchenko, A.S.; Balova, I.A.; Panov, M.S.; Kochemirovsky, V.A. In situ laser-induced synthesis of gas sensing microcomposites based on molybdenum and its oxides. *Compos. Part B Eng.* **2019**, *157*, 322–330. [CrossRef]

16. Smikhovskaia, A.V.; Panov, M.S.; Tumkin, I.I.; Khairullina, E.M.; Ermakov, S.S.; Balova, I.A.; Ryazantsev, M.N.; Kochemirovsky, V.A. In situ laser-induced codeposition of copper and different metals for fabrication of microcomposite sensor-active materials. *Anal. Chim. Acta* **2018**, *1044*, 138–146. [CrossRef]

17. Smikhovskaia, A.V.; Andrianov, V.S.; Khairullina, E.M.; Lebedev, D.V.; Ryazantsev, M.N.; Panov, M.S.; Tumkin, I.I. In situ laser-induced synthesis of copper-silver microcomposite for enzyme-free D-glucose and L-alanine sensing. *Appl. Surf. Sci.* **2019**, *488*, 531–536. [CrossRef]

18. Panov, M.; Aliabev, I.; Khairullina, E.; Mironov, V.; Tumkin, I. Fabrication of nickel-gold microsensor using in situ laser-induced metal deposition technique. *J. Laser Micro Nanoeng.* **2019**, *14*, 266–269. [CrossRef]

19. Panov, M.S.; Vereshchagina, O.A.; Ermakov, S.S.; Tumkin, I.I.; Khairullina, E.M.; Skripkin, M.Y.; Mereshchenko, A.S.; Ryazantsev, M.N.; Kochemirovsky, V.A. Non-enzymatic sensors based on in situ laser-induced synthesis of copper-gold and gold nano-sized microstructures. *Talanta* **2017**, *167*, 201–207. [CrossRef] [PubMed]

20. Andriianov, V.S.; Mironov, V.S.; Smikhovskaia, A.V.; Khairullina, E.M.; Tumkin, I.I. Laser-induced synthesis of carbon-based electrode materials for non-enzymatic glucose detection. *Opt. Quantum Electron.* **2020**, *52*, 45. [CrossRef]

21. Shishov, A.; Gordeychuk, D.; Logunov, L.; Tumkin, I. High rate laser deposition of conductive copper microstructures from deep eutectic solvents. *Chem. Commun.* **2019**, *55*, 9626–9628. [CrossRef] [PubMed]

22. Gorski, W.; Kennedy, R.T. Electrocatalyst for non-enzymatic oxidation of glucose in neutral saline solutions. *J. Electroanal. Chem.* **1997**, *424*, 43–48. [CrossRef]

23. Park, S.; Boo, H.; Chung, T.D. Electrochemical non-enzymatic glucose sensors. *Anal. Chim. Acta* **2006**, *556*, 46–57. [CrossRef] [PubMed]

24. Toghill, K.E.; Compton, R.G. Electrochemical non-enzymatic glucose sensors: A perspective and an evaluation. *Int. J. Electrochem. Sci.* **2010**, *5*, 1246–1301.

25. Si, P.; Huang, Y.; Wang, T.; Ma, J. Nanomaterials for electrochemical non-enzymatic glucose biosensors. *RSC Adv.* **2013**, *3*, 3487–3502. [CrossRef]

26. Miao, X.; Yang, C.; Leung, C.H.; Ma, D.L. Application of iridium(III) complex in label-free and non-enzymatic electrochemical detection of hydrogen peroxide based on a novel "on-off-on" switch platform. *Sci. Rep.* **2016**, *6*, 25774. [CrossRef]

27. Holt-Hindle, P.; Nigro, S.; Asmussen, M.; Chen, A. Amperometric glucose sensor based on platinum-iridium nanomaterials. *Electrochem. Commun.* **2008**, *10*, 1438–1441. [CrossRef]

28. Dong, Q.; Song, D.; Huang, Y.; Xu, Z.; Chapman, J.H.; Willis, W.S.; Li, B.; Lei, Y. High-temperature annealing enabled iridium oxide nanofibers for both non-enzymatic glucose and solid-state pH sensing. *Electrochim. Acta* **2018**, *281*, 117–126. [CrossRef]

29. Glebov, E.M.; Plyusnin, V.F.; Grivin, V.P.; Krupoder, S.A.; Liskovskaya, T.I.; Danilovich, V.S. Photochemistry of copper(II) polyfluorocarboxylates and copper(II) acetate as their hydrocarbon analogues. *J. Photochem. Photobiol. A Chem.* **2000**, *133*, 177–183. [CrossRef]

30. Khan, M.; Bouet, G.; Vierling, F.; Meullemeestre, J.; Schwing, M.J. Formation of cobalt(II), nickel(II) and copper(II) chloro complexes in alcohols and the Irving-Williams order of stabilities. *Transit. Met. Chem.* **1996**, *21*, 231–234. [CrossRef]

31. El-Khoury, P.Z.; Tarnovsky, A.N.; Schapiro, I.; Ryazantsev, M.N.; Olivucci, M. Structure of the photochemical reaction path populated via promotion of CF_2I_2 into its first excited state. *J. Phys. Chem. A* **2009**, *113*, 10767–10771. [CrossRef] [PubMed]

32. Matsumoto, A. *Topics in Current Chemistry Editorial Board*; Springer: Cham, Switzerland, 2005; ISBN 9783540733461.

33. Xu, Y.; Zhang, B. Recent advances in porous Pt-based nanostructures: Synthesis and electrochemical applications. *Chem. Soc. Rev.* **2014**, *43*, 2439–2450. [CrossRef] [PubMed]

34. Dolinska, J.; Kannan, P.; Sobczak, J.W.; Opallo, M. Glucose Electrooxidation in Bimetallic Suspensions of Nanoparticles in Alkaline Media. *ChemElectroChem* **2015**, *2*, 1199–1205. [CrossRef]

35. Hong, J.W.; Kim, Y.; Kwon, Y.; Han, S.W. Noble-Metal Nanocrystals with Controlled Facets for Electrocatalysis. *Chem.-Asian J.* **2016**, *11*, 2224–2239. [CrossRef] [PubMed]

36. Li, N.; Li, Q.; Yuan, M.; Guo, X.; Zheng, S.; Pang, H. Synthesis of $Co_{0.5}Mn_{0.1}Ni_{0.4}C_2O_4 \cdot n\ H_2O$ Micropolyhedrons: Multimetal Synergy for High-Performance Glucose Oxidation Catalysis. *Chem.—Asian J.* **2019**, *14*, 2259–2265. [CrossRef] [PubMed]

37. Hwang, D.W.; Lee, S.; Seo, M.; Chung, T.D. Recent advances in electrochemical non-enzymatic glucose sensors—A review. *Anal. Chim. Acta* **2018**, *1033*, 1–34. [CrossRef]

38. Pasta, M.; La Mantia, F.; Cui, Y. Mechanism of glucose electrochemical oxidation on gold surface. *Electrochim. Acta* **2010**, *55*, 5561–5568. [CrossRef]

39. Kim, S.H.; Choi, J.B.; Nguyen, Q.N.; Lee, J.M.; Park, S.; Chung, T.D.; Byun, J.Y. Nanoporous platinum thin films synthesized by electrochemical dealloying for nonenzymatic glucose detection. *Phys. Chem. Chem. Phys.* **2013**, *15*, 5782–5787. [CrossRef]

40. Shen, J.; Dudik, L.; Liu, C.C. An iridium nanoparticles dispersed carbon based thick film electrochemical biosensor and its application for a single use, disposable glucose biosensor. *Sens. Actuators B Chem.* **2007**, *125*, 106–113. [CrossRef]

41. Fu, S.; Zhu, C.; Song, J.; Engelhard, M.; Xia, H.; Du, D.; Lin, Y. PdCuPt Nanocrystals with Multibranches for Enzyme-Free Glucose Detection. *ACS Appl. Mater. Interfaces* **2016**, *8*, 22196–22200. [CrossRef]

42. Hsu, C.W.; Su, F.C.; Peng, P.Y.; Young, H.T.; Liao, S.; Wang, G.J. Highly sensitive non-enzymatic electrochemical glucose biosensor using a photolithography fabricated micro/nano hybrid structured electrode. *Sens. Actuators B Chem.* **2016**, *230*, 559–565. [CrossRef]

43. Qiu, H.; Huang, X. Effects of Pt decoration on the electrocatalytic activity of nanoporous gold electrode toward glucose and its potential application for constructing a nonenzymatic glucose sensor. *J. Electroanal. Chem.* **2010**, *643*, 39–45. [CrossRef]

44. Singh, B.; Dempsey, E.; Dickinson, C.; Laffir, F. Inside/outside Pt nanoparticles decoration of functionalised carbon nanofibers ($Pt_{19.2}$/f-$CNF_{80.8}$) for sensitive non-enzymatic electrochemical glucose detection. *Analyst* **2012**, *137*, 1639–1648. [CrossRef] [PubMed]

45. Wang, J.; Zhao, D.; Xu, C. Nonenzymatic electrochemical sensor for glucose based on nanoporous platinum-gold alloy. *J. Nanosci. Nanotechnol.* **2016**, *16*, 7145–7150. [CrossRef]

MDPI

St. Alban-Anlage 66

4052 Basel

Switzerland

Tel. +41 61 683 77 34

Fax +41 61 302 89 18

www.mdpi.com

Materials Editorial Office

E-mail: materials@mdpi.com

www.mdpi.com/journal/materials